NATIONAL COOPERATIVE HIGHWAY RESEARCH PROGRAM

NCHRP SYNTHESIS 331

Statewide Highway Letting Program Management

A Synthesis of Highway Practice

CONSULTANTS
STUART D. ANDERSON
Texas Transportation Institute
and
BYRON C. BLASCHKE
College Station, Texas

TOPIC PANEL

WILLIAM L. HEBERT, *New York State Department of Transportation*
FREDERICK D. HEJL, *Transportation Research Board*
DAVID MAYO, *Arkansas Highway and Transportation Department*
SCOTT NICHOLS, *Texas Department of Transportation*
CLIFF SCHEXNAYDER, *Arizona State University*
STEVE TEAGUE, *Arkansas Highway and Transportation Department*
JOHN O'FALLON, *Federal Highway Administration (Liaison)*
EDWIN ONKONKWO, *Federal Highway Administration (Liaison)*

SUBJECT AREAS
Planning and Administration and Materials and Construction

Research Sponsored by the American Association of State Highway and Transportation Officials in Cooperation with the Federal Highway Administration

TRANSPORTATION RESEARCH BOARD

WASHINGTON, D.C.
2004
www.TRB.org

NATIONAL COOPERATIVE HIGHWAY RESEARCH PROGRAM

Systematic, well-designed research provides the most effective approach to the solution of many problems facing highway administrators and engineers. Often, highway problems are of local interest and can best be studied by highway departments individually or in cooperation with their state universities and others. However, the accelerating growth of highway transportation develops increasingly complex problems of wide interest to highway authorities. These problems are best studied through a coordinated program of cooperative research.

In recognition of these needs, the highway administrators of the American Association of State Highway and Transportation Officials initiated in 1962 an objective national highway research program employing modern scientific techniques. This program is supported on a continuing basis by funds from participating member states of the Association and it receives the full cooperation and support of the Federal Highway Administration, United States Department of Transportation.

The Transportation Research Board of the National Research Council was requested by the Association to administer the research program because of the Board's recognized objectivity and understanding of modern research practices. The Board is uniquely suited for this purpose as it maintains an extensive committee structure from which authorities on any highway transportation subject may be drawn; it possesses avenues of communication and cooperation with federal, state, and local governmental agencies, universities, and industry; its relationship to the National Research Council is an insurance of objectivity; it maintains a full-time research correlation staff of specialists in highway transportation matters to bring the findings of research directly to those who are in a position to use them.

The program is developed on the basis of research needs identified by chief administrators of the highway and transportation departments and by committees of AASHTO. Each year, specific areas of research needs to be included in the program are proposed to the National Research Council and the Board by the American Association of State Highway and Transportation Officials. Research projects to fulfill these needs are defined by the Board, and qualified research agencies are selected from those that have submitted proposals. Administration and surveillance of research contracts are the responsibilities of the National Research Council and the Transportation Research Board.

The needs for highway research are many, and the National Cooperative Highway Research Program can make significant contributions to the solution of highway transportation problems of mutual concern to many responsible groups. The program, however, is intended to complement rather than to substitute for or duplicate other highway research programs.

NCHRP SYNTHESIS 331

Project 20-5 FY 2000 (Topic 33-09)
ISSN 0547-5570
ISBN 0-309-07008-2
Library of Congress Control No. 2004105196

Price $15.00

NOTICE

Published reports of the

NATIONAL COOPERATIVE HIGHWAY RESEARCH PROGRAM

are available from:

Transportation Research Board
Business Office
500 Fifth Street
Washington, D.C. 20001

and can be ordered through the Internet at:

http://www.national-academies.org/trb/bookstore

Printed in the United States of America

THE NATIONAL ACADEMIES

Advisers to the Nation on Science, Engineering, and Medicine

The **National Academy of Sciences** is a private, nonprofit, self-perpetuating society of distinguished scholars engaged in scientific and engineering research, dedicated to the furtherance of science and technology and to their use for the general welfare. On the authority of the charter granted to it by the Congress in 1863, the Academy has a mandate that requires it to advise the federal government on scientific and technical matters. Dr. Bruce M. Alberts is president of the National Academy of Sciences.

The **National Academy of Engineering** was established in 1964, under the charter of the National Academy of Sciences, as a parallel organization of outstanding engineers. It is autonomous in its administration and in the selection of its members, sharing with the National Academy of Sciences the responsibility for advising the federal government. The National Academy of Engineering also sponsors engineering programs aimed at meeting national needs, encourages education and research, and recognizes the superior achievements of engineers. Dr. William A. Wulf is president of the National Academy of Engineering.

The **Institute of Medicine** was established in 1970 by the National Academy of Sciences to secure the services of eminent members of appropriate professions in the examination of policy matters pertaining to the health of the public. The Institute acts under the responsibility given to the National Academy of Sciences by its congressional charter to be an adviser to the federal government and, on its own initiative, to identify issues of medical care, research, and education. Dr. Harvey V. Fineberg is president of the Institute of Medicine.

The **National Research Council** was organized by the National Academy of Sciences in 1916 to associate the broad community of science and technology with the Academy's purposes of furthering knowledge and advising the federal government. Functioning in accordance with general policies determined by the Academy, the Council has become the principal operating agency of both the National Academy of Sciences and the National Academy of Engineering in providing services to the government, the public, and the scientific and engineering communities. The Council is administered jointly by both Academies and the Institute of Medicine. Dr. Bruce M. Alberts and Dr. William A. Wulf are chair and vice chair, respectively, of the National Research Council.

The **Transportation Research Board** is a division of the National Research Council, which serves the National Academy of Sciences and the National Academy of Engineering. The Board's mission is to promote innovation and progress in transportation through research. In an objective and interdisciplinary setting, the Board facilitates the sharing of information on transportation practice and policy by researchers and practitioners; stimulates research and offers research management services that promote technical excellence; provides expert advice on transportation policy and programs; and disseminates research results broadly and encourages their implementation. The Board's varied activities annually engage more than 5,000 engineers, scientists, and other transportation researchers and practitioners from the public and private sectors and academia, all of whom contribute their expertise in the public interest. The program is supported by state transportation departments, federal agencies including the component administrations of the U.S. Department of Transportation, and other organizations and individuals interested in the development of transportation. **www.TRB.org**

www.national-academies.org

FOREWORD

By Staff
Transportation Research Board

Highway administrators, engineers, and researchers often face problems for which information already exists, either in documented form or as undocumented experience and practice. This information may be fragmented, scattered, and unevaluated. As a consequence, full knowledge of what has been learned about a problem may not be brought to bear on its solution. Costly research findings may go unused, valuable experience may be overlooked, and due consideration may not be given to recommended practices for solving or alleviating the problem.

Information exists on nearly every subject of concern to highway administrators and engineers. Much of it derives from research or from the work of practitioners faced with problems in their day-to-day work. To provide a systematic means for assembling and evaluating such useful information and to make it available to the entire highway community, the American Association of State Highway and Transportation Officials—through the mechanism of the National Cooperative Highway Research Program—authorized the Transportation Research Board to undertake a continuing study. This study, NCHRP Project 20-5, "Synthesis of Information Related to Highway Problems," searches out and synthesizes useful knowledge from all available sources and prepares concise, documented reports on specific topics. Reports from this endeavor constitute an NCHRP report series, *Synthesis of Highway Practice*.

The synthesis series reports on current knowledge and practice, in a compact format, without the detailed directions usually found in handbooks or design manuals. Each report in the series provides a compendium of the best knowledge available on those measures found to be the most successful in resolving specific problems.

PREFACE

This report of the Transportation Research Board is a study of the current practices for identifying, measuring, and articulating the public benefits of highway system maintenance and operation, and of communicating those benefits that are understandable and meaningful to stakeholders—road users, elected officials, and others who have an interest in the system's performance. It includes information on the difficulties public agencies encounter in explaining these benefits.

This synthesis report included a review of published literature on the measurement and communication of maintenance benefits, a formal survey of state transportation agencies, and informal interviews and discussions with a range of individuals engaged in highway system management.

A panel of experts in the subject area guided the work of organizing and evaluating the collected data and reviewed the final synthesis report. A consultant was engaged to collect and synthesize the information and to write this report. Both the consultant and the members of the oversight panel are acknowledged on the title page. This synthesis is an immediately useful document that records the practices that were acceptable within the limitations of the knowledge available at the time of its preparation. As progress in research and practice continues, new knowledge will be added to that now at hand.

CONTENTS

ACKNOWLEDGMENTS

Stuart D. Anderson, Texas Transportation Institute, and Byron C. Blaschke, College Station, Texas, were responsible for collection of the data and preparation of the report.

Valuable assistance in the preparation of this synthesis was provided by the Topic Panel, consisting of William L. Hebert, Associate Transportation Analyst, New York State Department of Transportation; Frederick D. Hejl, Senior Program Officer, Transportation Research Board; David Mayo, Staff Project Development Engineer, Arkansas Highway and Transportation Department; Scott Nichols, Manager, Contract Letting and Processing Branch, Texas Department of Transportation; John O'Fallon, Bridge Research Engineer, Turner–Fairbank Highway Research Center, Federal Highway Administration; Edwin Onkonkwo, Federal Highway Administration, HIPA-30; Cliff Schexnayder, Arizona State University; and Steve Teague, Assistant to the Director for Program Management, Arkansas Highway and Transportation Department.

This study was managed by Stephen F. Maher, P.E., and Jon Williams, Managers, Synthesis Studies, who worked with the consultant, the Topic Panel, and the Project 20-5 Committee in the development and review of the report. Assistance in project scope development was provided by Donna Vlasak, Senior Program Officer. Don Tippman was responsible for editing and production. Cheryl Keith assisted in meeting logistics and distribution of the questionnaire and draft reports.

Crawford F. Jencks, Manager, National Cooperative Highway Research Program, assisted the NCHRP 20-5 Committee and the Synthesis staff.

Information on current practice was provided by many highway and transportation agencies. Their cooperation and assistance are appreciated.

STATEWIDE HIGHWAY LETTING PROGRAM MANAGEMENT

SUMMARY

One goal of state highway agencies (SHAs) is to maintain, upgrade, and improve the highway systems within the state. To achieve this goal, SHAs must identify highway transportation needs, prioritize these needs, and then address the needs through individual projects. The list of needs is large and the number of projects is correspondingly large. The set of projects that are in advanced stages of design and that have a target date for a construction bid is known as a "letting program." The development and management of a letting program that can bring projects to fruition within the available funds, and that satisfy many different constituents, is a significant challenge for each SHA.

A review of the literature was conducted pertaining to letting program management, with a specific focus on processes and techniques that describe statewide highway letting program management. A survey questionnaire was administered that focused on different aspects of SHA letting programs. Selected telephone interviews were conducted with SHAs to explore in more detail the approach taken to develop and then to manage the letting program. Twenty-eight agencies responded to the survey, and five agencies participated in the interviews.

A formalized and structured approach to letting program management was not found in the literature nor specifically captured through the survey of SHAs. Thus, a generic picture that best represents components of SHA highway letting program management was developed. Key information required for these components is identified, as well as the interaction between components and various agencies or groups involved in letting program management.

Statewide highway letting program development and management is complex. There are many agencies involved in letting program management. Federal and state requirements and regulations influence how this process is conducted. Within the SHA, many different organizational units are involved in statewide highway letting program management.

The first year of the Statewide Transportation Improvement Program (STIP) is the primary determinant of which projects are included in the letting program or "pool of projects." Many entities, including the general public, provide input about which projects are to be included in the STIP and the priority of these projects. Key factors that drive the priority of projects in the STIP include safety, level of traffic (e.g., average daily traffic), consistency with long-range plans, cost-effectiveness, and condition of the existing facility. Development of the STIP is closely tied to the project development process with respect to the status of design completion. However, there are many different ways in which SHAs select specific projects for inclusion in their letting programs. There appear to be some dominant factors that influence this decision such as delivery status of the project, and project priority established by SHA districts, regions, and divisions, with input from the FHWA.

Most SHAs follow the generic process developed for this synthesis in some form, with the approach varying depending on state requirements. Matching funds to available projects is a key requirement to move projects onto letting schedules. Federal-aid projects must comply with FHWA requirements. States must also closely monitor current cash flow and ensure that sufficient funds are available before moving projects into letting schedules. Other constraints can also influence which projects are included in the letting schedule, such as SHA policy, the geographical distribution of projects to districts or regions, and the priority set for the project based on the letting program and the STIP. The projected completion date and final approval of plans, specifications, and estimates confirm the letting date on the schedule.

Letting schedules are constantly changing for many reasons. Managing the impact of volatility in letting schedules is critical to using available funds and to ensuring that projects are awarded for construction. Many factors can influence the project construction contract. These factors, when considered during bid analysis, may result in delaying or not awarding the contract. Finally, data sharing about letting programs among SHAs was found to be a low-priority item.

Statewide highway letting program management practices are not well documented. The complexity and breadth of letting program management make it difficult for any one person or unit within the SHA to fully understand the entire process. Finally, there appears to be a lack of a single documented and comprehensive approach to developing and managing the letting program. There are few comprehensive tools and techniques available to SHAs that specifically focus on statewide highway letting program management.

CHAPTER ONE

INTRODUCTION

BACKGROUND

Maintaining, upgrading, and improving the highway system within the state is one goal of state highway agencies (SHAs). To achieve this goal, SHAs must identify highway transportation needs, prioritize these needs, and then address the needs through individual projects. The list of needs is enormous, with the number of projects correspondingly enormous. In addition, like most organizations, SHAs have limited resources and funds to allocate to the development of these projects. The development and management of a program that can bring projects to fruition, within the available funds, and that can satisfy the many different constituents, is a significant challenge for each SHA.

SHAs have processes to identify specific needs for highway improvements. Through these processes, needs are prioritized and costs estimated as required to translate the need into a viable project. The estimated cost for selected projects must be congruent with forecasted funds. When funds are foreseen for a viable project, the project is programmed and authorized for further development. This authorization initiates advanced planning and preliminary design, including environmental clearance. The project then moves through advanced planning and preliminary design to final design. When final design is complete, the environmental clearance issued, and the right-of-way (ROW) acquired, the project is let for bid. If a satisfactory bid is received, the project is awarded, a contract signed, and construction commences.

During project development the SHA begins to establish a schedule to identify when projects will be contracted for construction. This action results in a letting program. The letting program can be described in terms of a "pool of projects" that are considered to be in an advanced stage of design, when plans, specifications, and estimates (PS&Es) are substantially or 100% complete. At this stage, the project has a target date scheduled for letting (month and year).

SHAs have processes for developing their letting programs. These processes are influenced by federal and state requirements, input provided by various agencies and the public, and capital budgets for funding projects. Similarily, SHAs have processes, with varying degrees of structure and complexity, to manage their letting programs. These processes require the use of tools and techniques to schedule lettings and award transportation projects in compliance with established criteria. However, the letting program process is subjected to change as a result of numerous circumstances, both internal and external. Program modifications can be caused by funding, design, or schedule changes. Understanding the causes behind this potential for change and how change is managed with respect to the letting process is important.

The development and management of statewide highway letting programs is a critical component of successful project delivery. Successful project delivery will support achievement of local, state, and national transportation goals. This synthesis discusses the processes, techniques, tools, and other critical requirements used by SHAs for statewide highway letting program management.

SYNTHESIS OBJECTIVES

The purpose of this synthesis is to summarize available information on statewide highway letting programs and document how SHAs manage these programs. Therefore, this synthesis identifies the processes involved with statewide highway letting program management. There are five objectives of this synthesis study:

- Identify elements, approaches, tools, and techniques used to develop letting programs;
- Identify elements, approaches, tools, and techniques considered in the management of letting programs;
- Evaluate how change influences letting schedules;
- Identify the impact of issues relevant to contract award that may influence the outcomes of specific lettings; and
- Assess SHA interest in data sharing initiatives relative to letting program management.

APPROACH AND METHODOLOGY

The scope of this synthesis covers the preparation and administration of the SHA letting program from its inception in planning and programming through construction contract award. Based on the synthesis purpose and objectives, three main sources of information were used to develop the synthesis material:

1. A review of the literature pertaining to letting program management was conducted with a specific fo-

cus on processes and techniques that describe statewide highway letting program management.
2. A survey questionnaire was administered to SHAs that focused on different aspects of SHA letting programs.
3. Selected interviews with SHAs were conducted to explore in detail the approaches taken to develop and manage the letting program.

Literature Review

A review of the literature relative to letting program management was conducted. A primary source was the Transportation Research Information Services (TRIS). Other databases were also explored, including some Internet sites. In general, the literature review provided very little information related to letting programs. Some literature was found on specific elements of letting programs; however, this information was minimal. Much of the literature studied focused on information developed by the FHWA to guide SHAs in developing and managing certain aspects of their letting programs. Analyses of these few references provided some insights into the general nature of the letting program process. There were no references that focused on this subject matter comprehensively.

Survey Questionnaire

A survey was designed to collect information on various aspects of letting programs. Five sections were identified for the survey questionnaire, in accordance with initial study objectives:

- Project Development and Letting Programs—identified how the letting program fits within a SHA's project development process (PDP). Furthermore, this section determined differences between the administration of state and federally funded projects and how SHAs ensure that funds are fully utilized to support the letting program.
- Letting Program—defined the elements of letting programs and, specifically, processes used to develop letting programs, how specific projects are selected for inclusion in the letting program, frequency of project lettings, how projects are selected for specific lettings, the impact of state and federal laws on letting program steps, and the role of other agencies involved in the letting program.
- Volatility in Letting Program—identified major factors that may change the letting program, probable causes of these changes, management actions taken to mitigate the impact that results from changes, and the impact that change has on the overall letting program.
- Contract Award Considerations—covered a number of issues that may influence the award of a contract for a specific letting, such as criteria for rejecting bids, analyzing bids that appear unbalanced, identifying collusion, determining bid responsiveness, policies for handling errors, use of electronic bidding, use of pre-bid conferences, frequency of awarding projects to the low bidder, and frequency of 100% clearance for utilities, ROW, and permits before contract award.
- Data Sharing—considered the types of data collected with respect to the SHA letting program, the types of data that an SHA would like to collect, the extent letting program data are shared with or received from other agencies, the importance of data sharing in this area, and finally the use of software to support different aspects of statewide highway letting program management.

The survey questionnaire is presented in Appendix A.

The questions were structured primarily for open-ended responses. This type of questioning was necessary to ensure that an adequate understanding could be gained concerning SHA approaches to letting program management. In addition, some questions requested additional supporting information in the form of hard copy or website addresses, where documentation related to letting program management could be found. Appendix B provides SHA website references that illustrate different components of the letting program process.

Survey Interviews

After analysis of the information collected, more specific and targeted data collection was required to better understand and clarify approaches to management of statewide highway letting programs. An interview protocol was developed and used as a guide to conducting a small number of telephone interviews. The interview protocol included generic flowcharts that described the development and management of letting programs. Questions were asked specifically in relation to the generic flowcharts. The interview protocol is provided in Appendix C.

Survey Responses

The questionnaire was distributed to state highway agencies in all 50 states, the District of Columbia, and Puerto Rico, and to the 13 provincial transportation agencies in Canada. A total of 28 responses were received, all representing U.S. SHAs (Table 1). The data collected were summarized and analyzed, with the results of this analysis discussed in the appropriate chapters.

Telephone interviews were requested of nine SHAs, each of which had completed a questionnaire. Five of the nine SHAs actually participated in the interviews (see Table 1). They represented a cross section of agencies with small, medium-sized, and large construction programs based on annual construction volume (dollars). The data from the interviews were analyzed and the results incorporated into the appropriate chapters.

TABLE 1
STATE HIGHWAY AGENCIES PARTICIPATING IN DATA COLLECTION

Questionnaire Response	Interview Response
Arizona	
Arkansas	Arkansas
California	
Connecticut	Connecticut
Delaware	
Florida	
Georgia	
Hawaii	
Idaho	
Iowa	
Kansas	Kansas
Louisiana	
Maine	
Maryland	
Minnesota	
Nebraska	
Nevada	
New Hampshire	
New Jersey	
New York	New York
North Dakota	
Ohio	
Pennsylvania	
Texas	Texas
Virginia	
Washington	
Wisconsin	
Wyoming	

Definitions

Certain terms that require definition are used throughout this synthesis. These terms are not necessarily common nor are they frequently used terms in SHA practice. Rather, they are used in this synthesis to define specific features of statewide highway letting program development and management. The key terms are defined as follows:

- *Project development*—a series of processes (e.g., planning, programming, design, and construction) that convert a highway transportation need into a completed facility that satisfies the need.
- *Viable project*—a scope and concept of work with identified limits, meeting a transportation need(s), and consistent with long-range plans.
- *Letting program process*—a series of steps that uses the products of the planning and programming functions as a basis for authorizing and controlling the stages of project development and for establishing the time schedules for letting projects. The process also incorporates the management of the timetable for the flow of projects from initial development authorization through letting.
- *Letting program or pool of projects*—a set of projects that are in the advanced stages of design, each with a target time for construction bid identified (month and year).
- *Letting schedule*—a document that lists projects and specific dates on which the projects will be bid for construction (month, day, and year). Typically, it includes projects that will be let in a period of 1 year or less.
- *Letting*—a function that includes advertisement of the proposed construction projects, receipt of bids, and the opening and reading of the bids in a public setting.

ORGANIZATION OF THE SYNTHESIS

The report contains chapters presenting various aspects of letting program management, as practiced by SHAs. The first chapter introduces the subject area and presents the study objectives, study approach, and methodology. The second chapter provides an overview of the PDP and the development and management of letting programs. The interaction between the PDP and the letting program process is discussed. Chapter three describes specifically the development of the letting program for an SHA. The main focus of this chapter is related to how SHAs form the pool of projects that constitutes an SHA letting program. Chapter four discusses the management of the letting program, specifically, approaches to developing letting schedules, changes in program letting schedules, and how change is managed. Specific information on an important aspect of the letting process, construction contract award considerations, is described. Finally, data sharing in the area of program letting management is reviewed. Chapter five consists of concluding remarks drawn from and substantiated in the preceding text of the synthesis report.

CHAPTER TWO

PROJECT DEVELOPMENT AND LETTING PROGRAM

INTRODUCTION

Because the PDP involves numerous activities, several disciplines, and many different interactions, it can be very complex. The process is guided by federal and state regulations, which can be accomplished quickly for some projects or can take many years for others. A key step early in the PDP is to identify needs for highway improvement, prioritize these needs, and determine the funding required for meeting the needs. These needs are referred to as "viable projects" and are incorporated into a multiyear, long-range plan of potential projects. Based on projected funds, viable projects are programmed and authorized for further development. Authorized projects are normally assigned a project identifier and entered into a project development tracking system. Authorized projects move through advanced planning and preliminary design, including environmental clearance to final design and finally, when design is complete and ROW is acquired, to bid letting. The project is normally awarded, if the bid is responsive, and then construction commences. All SHAs have PDPs that generally follow these stages. At some point in the PDP the SHA establishes a letting program. As previously mentioned, the letting program constitutes a pool of projects. When the final design of a project in the pool is close to completion, the project can be scheduled for letting on a specific month and day.

As stated earlier, there are two components of a letting program: (1) development of the program and (2) management of the program. SHAs have different approaches to the development and management of their statewide highway letting programs. Both development and management of the letting program are influenced by federal and state requirements, funds available, and many other factors. This chapter provides an overview of the processes followed to develop and manage the letting program. These processes are linked closely with the PDP.

PROJECT DEVELOPMENT PROCESS

The PDP is described in several ways in the literature. Saag (1999) has identified six different components: (1) planning, (2) project development, (3) mitigation, (4) right-of-way, (5) design, and (6) construction. Earlier, Anderson and Fisher (1997) described the PDP as having three main components: (1) planning, (2) design, and (3) construction. In Anderson and Fisher's characterization, planning is described according to project definition and conceptual plan development. Design comprises preliminary design, PS&E development, and final design. Finally, construction is described in terms of pre-construction, construction, and post-construction. The FHWA describes the process in two phases: planning and project development ("Contract Administration Core . . ." 2001). The planning process described by the FHWA focuses on planning and programming, and developing long-range plans based on transportation needs and then short-range plans focusing on specific projects. Although each of these three depictions describes the PDP by using different terms, upon inspection of the details of each, it can be seen that the processes cover the same basic stages.

Each SHA has its own version of the PDP. The terms used to describe these processes vary by state. For example, Arizona has a seven-stage PDP that includes identification, scoping, programming, designing, advertising, awarding, and construction. Pennsylvania's characterization of its PDP is described as planning, prioritization and programming, preliminary design, final design, and construction. Some SHAs, in describing their PDPs, included stages similar to those in the FHWA planning process. For example, Arkansas described their PDP as project need identification, job programmed, job added to the 3-year Statewide Transportation Improvement Program (STIP), design and environmental clearance, ROW acquisition, and construction contract award. Although the terms used by SHAs to describe the PDP are frequently different, the overall focus is the same.

Owing to the various different descriptions, and for purposes of this synthesis, the PDP will be described in terms of the following seven basic stages: (1) planning, (2) programming, (3) advanced planning and preliminary design, (4) final design, (5) letting, (6) award, and (7) construction (see Figure 1). Table 2 shows some basic activities involved with each stage.

LETTING PROGRAM PROCESS

There exists a limited amount of information describing the management of statewide highway letting programs. The FHWA describes the output of planning and programming at two levels: planning, which produces a set of viable projects that address long-range statewide and metropolitan highway improvement needs and, at the second

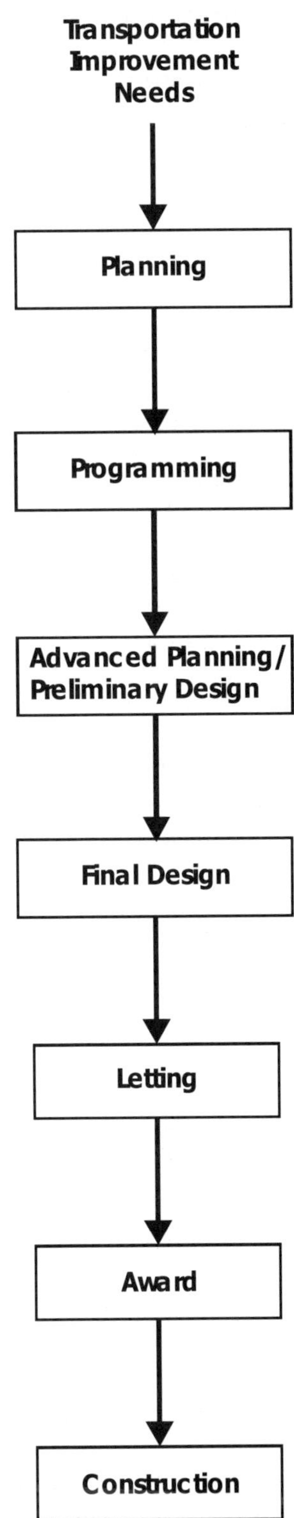

FIGURE 1 Typical stages in the project development process.

level, where the needs in the long-range plan are translated into authorized projects included in a 3-year STIP. According to the FHWA, projects are initiated based on the STIP, but there can be other sources of initiation as well. In the FHWA description, the project flows through the development process with no further depiction or discussion of letting program management ("Contract Administration Core . . ." 2001).

TABLE 2
PROJECT DEVELOPMENT STAGES AND ACTIVITIES

Stages of Project Development Process	Typical Activities
Planning	Purpose and need, improvement or requirement studies, environmental considerations, and interagency coordination.
Programming	Environmental determination, schematic development, public hearings, ROW plan, and project funding authorization.
Advanced Planning/ Preliminary Design	ROW development, environmental clearance, design criteria and parameters, surveys/utility locations/drainage, preliminary schematics such as alternative selections, geometric alignments, and bridge layouts.
Final Design	ROW acquisition, PS&E development—pavement and bridge design, traffic control plans, utility drawings, hydraulic studies/drainage design, and final cost estimates.
Letting	Prepare contract documents, advertise for bid, pre-bid conference, and receive and analyze bids.
Award	Determine lowest responsive bidder, initiate contract.
Construction	Mobilization, inspection and materials testing, contract administration, and traffic control, bridge, pavement, and drainage construction.

Notes: ROW = right-of-way; PS&E = plans, specifications, and estimates.
Sources: Anderson and Fisher 1997; Saag 1999.

A specific question survey concerned the process that an SHA follows to develop its letting program. This question was answered in a variety of ways and from many different perspectives. One finding, after analyzing the answers, is that none of the SHAs responding to the questionnaire have a well-defined and comprehensively documented letting program process. Each answer provided by an SHA represented some component of the process, with some answers providing a more comprehensive description of a component(s) than others. Thus, based on what literature was available and results from this question, a macro-level characterization of a letting program process was developed and described by two main components: development of the letting program and management of the letting program (Figure 2). The validity of this characterization was confirmed through the five telephone interviews.

The development of the letting program, for purposes of this synthesis, begins with an authorized program of projects, as shown in the upper portion of Figure 2. This program usually covers a period of 12 years or less. The projects in the program may be generally prioritized with target letting dates. On the basis of the authorized program,

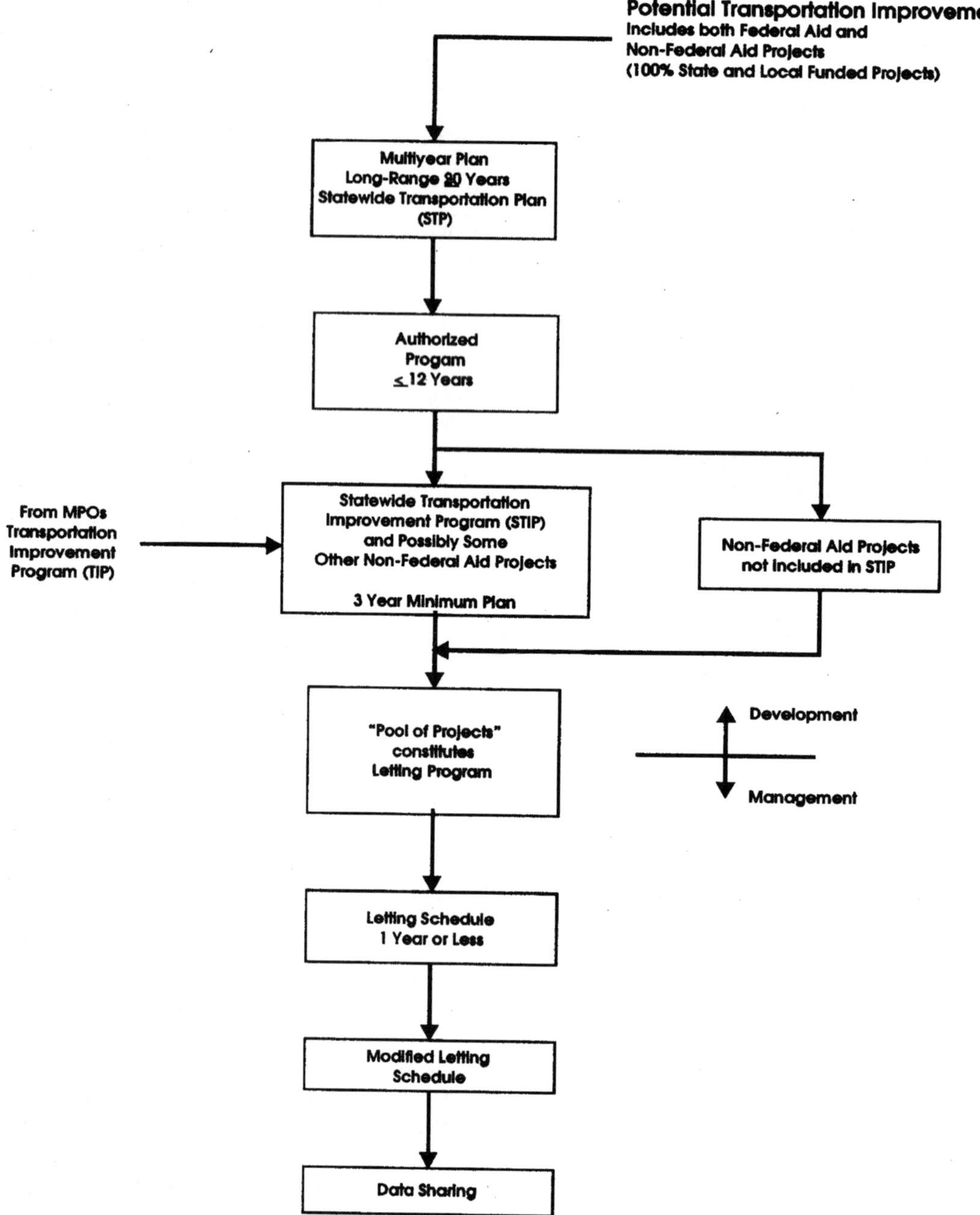

FIGURE 2 Components of the letting program process.

the SHA formulates an STIP. The STIP includes projects from the statewide transportation plan [multiyear, long-range plan (at least 20 years)] and projects from metropolitan planning organizations (MPOs) as identified in an MPO's Transportation Improvement Programs (TIPs). The STIP covers at least a 3-year window of projects as required for federal-aid projects. Large, complex projects in the STIP are typically in the final design development stage of the PDP. However, the STIP may also include smaller and less complex projects that are in the earlier stages of the PDP (i.e., in the advanced planning and preliminary design stages). Non-federal-aid projects authorized by the state may also be included in this STIP. The pool of projects that this synthesis describes as the letting program comes from the STIP and can include other non-federal-aid projects that may not be included in the STIP.

Once the pool of projects or letting program is established, the management of the letting program commences, as shown in the lower portion of Figure 2. Projects move from the pool of projects into a letting schedule that covers a time frame of typically 1 year or less. Projects in the pool are placed on a letting schedule based on design status, funding availability, and other constraints (e.g., policy and administrative considerations and environmental clearance). These projects have a specific letting date. However, the letting program process can experience changes or changing conditions can result in modification of letting schedules. As a result of change, some projects may return to the pool of projects, and substitute projects are often added to the letting schedule. Most projects let for bid are awarded. However, some projects that have been bid and not awarded may be rebid at a later letting or returned to the pool of projects based on certain award considerations (e.g., nonresponsive bid or bid prices substantially higher or lower than the state agency's engineer's estimate determined at the time of PS&E completion).

INTERACTION OF PROJECT DEVELOPMENT PROCESS AND LETTING PROGRAM MANAGEMENT

The PDP and letting program management interact. This interaction is complex and is characterized by many inputs and outputs, and factors that influence the generation of these inputs and outputs. Figure 3 combines and expands Figures 1 and 2 and shows the interaction based on interface points between the PDP and the process for developing and managing the letting program. The letting program process is displayed in the left side of Figure 3, whereas the PDP is shown on the right side. A dashed line is used to separate these two processes.

The involvement of key agencies and the flow of key information are incorporated into the components of each process. Figure 3 was developed using literature from the FHWA; literature from SHAs, as provided through SHA websites or documents returned with the survey; questionnaire responses; and the author's experience. Telephone interviews were also conducted to confirm the information shown in this figure. The interviews were based on questions specifically related to components of the process for developing and managing the letting program, as shown in Figure 3.

An interview process was necessary for this synthesis because the data and information retrieved from questionnaire responses gave only snapshots of process elements. No single SHA provided a complete process picture representing statewide letting program management. However, one question from the survey did help identify where the letting program process starts and where it is completed. This demarcation has aided in portraying graphically the development and management of the letting program process. The specific question focused on defining at which phase in the SHA PDP a project was included in the SHA letting program.

By using the seven stages in the generic PDP shown in Figure 1, and the SHA PDP description provided from the questionnaire response, the PDP phase in which an SHA incorporates a project into the SHA letting program was matched with a similar stage in the generic PDP. From this analysis, 12 SHAs considered the start of the letting program to be at the end of programming or at the transition to advanced planning and preliminary design. Therefore, this stage could be considered the start of the letting program. Nine SHAs stated that a project is included in the letting program during final design stages, and three SHAs viewed the letting program as beginning when a project is ready to be advertised for letting. This range of answers indicated that letting program management could perhaps start as early as when projects are programmed. This PDP stage creates the "Authorized Program" as shown in Figure 3. Letting program management is complete when a project is awarded for construction. These findings and interpretation demonstrate some consistency with the FHWA approach to planning and programming projects ("Contract Administration Core . . ." 2001). Figure 3, therefore, represents a synthesis of various SHA practices based on the interpretation of the information provided by survey respondents and confirmed through subsequent interviews with five SHAs.

Another way to view the process of developing and managing a letting program is to consider the process as the flow of projects through a piping system. This pipeline system analogy is shown in the Figure 4 schematic. Viable projects enter the pipeline system based on a long-range plan. These viable projects may be filtered by project and funding categories. For projects to advance through the system, at various points they must be authorized, developed, and approved to a final design. Certain criteria influence the rate of project flow through the system. At various points, valves are turned "on and off" to control the project flow rate. These valves represent critical elements, constraints, decision points, or other considerations that must be addressed for the project to proceed through the system. The various approaches, tools, and techniques that SHAs employ to manage the flow of projects through their system are described in subsequent chapters.

SUMMARY

Statewide highway letting program management is complex. There are many different agencies, such as MPOs and the FHWA, that provide input during letting program management. Federal and state requirements and regulations influence how this process is conducted. Funding significantly influences the development and management of letting programs. Within the SHA, many different organizational units are involved in managing the letting program. A formalized and structured approach to letting program management was not found in the literature nor specifically captured through the survey of SHAs. Thus, a generic

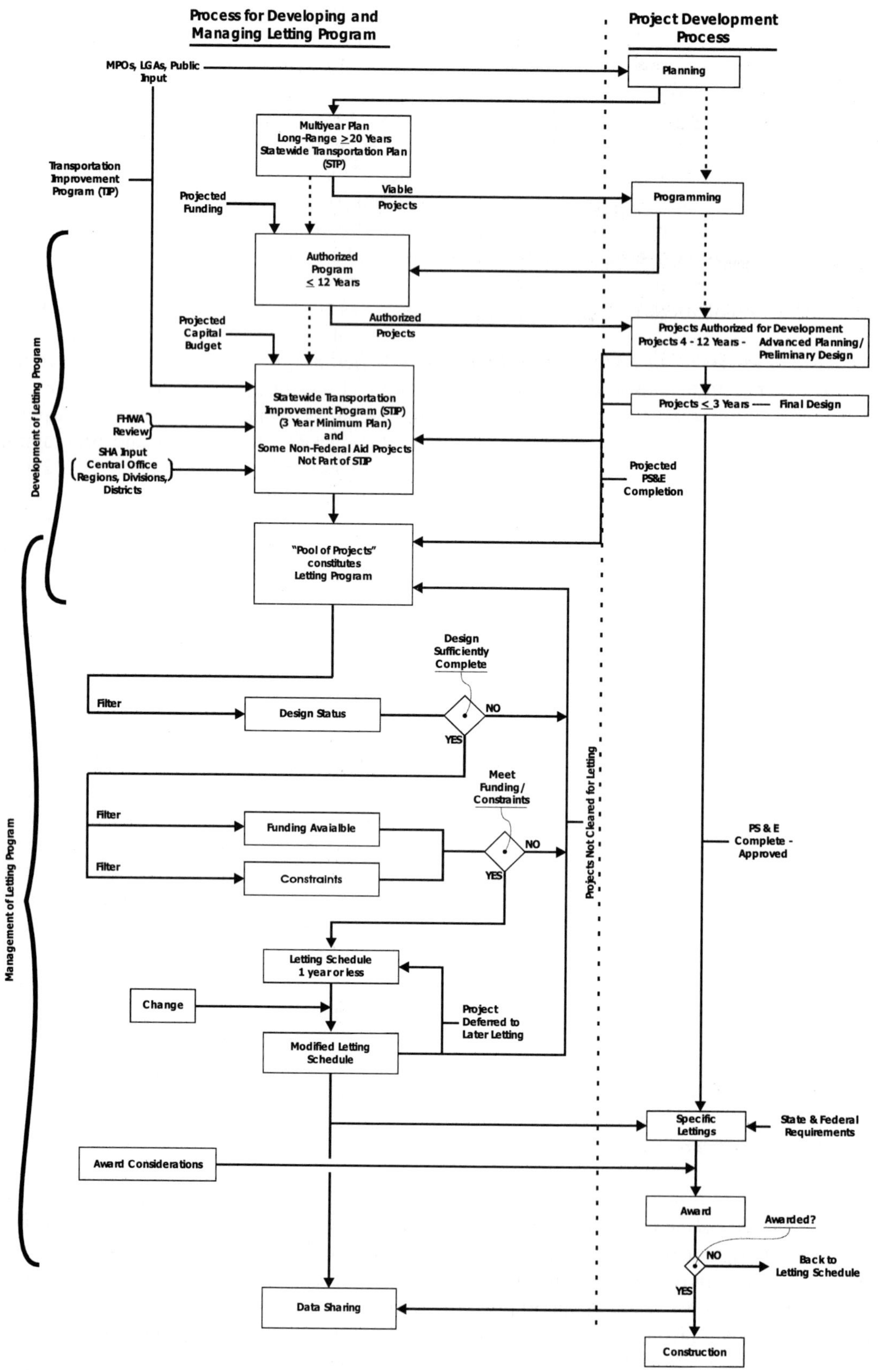

FIGURE 3 Interaction of project development process and letting program process.

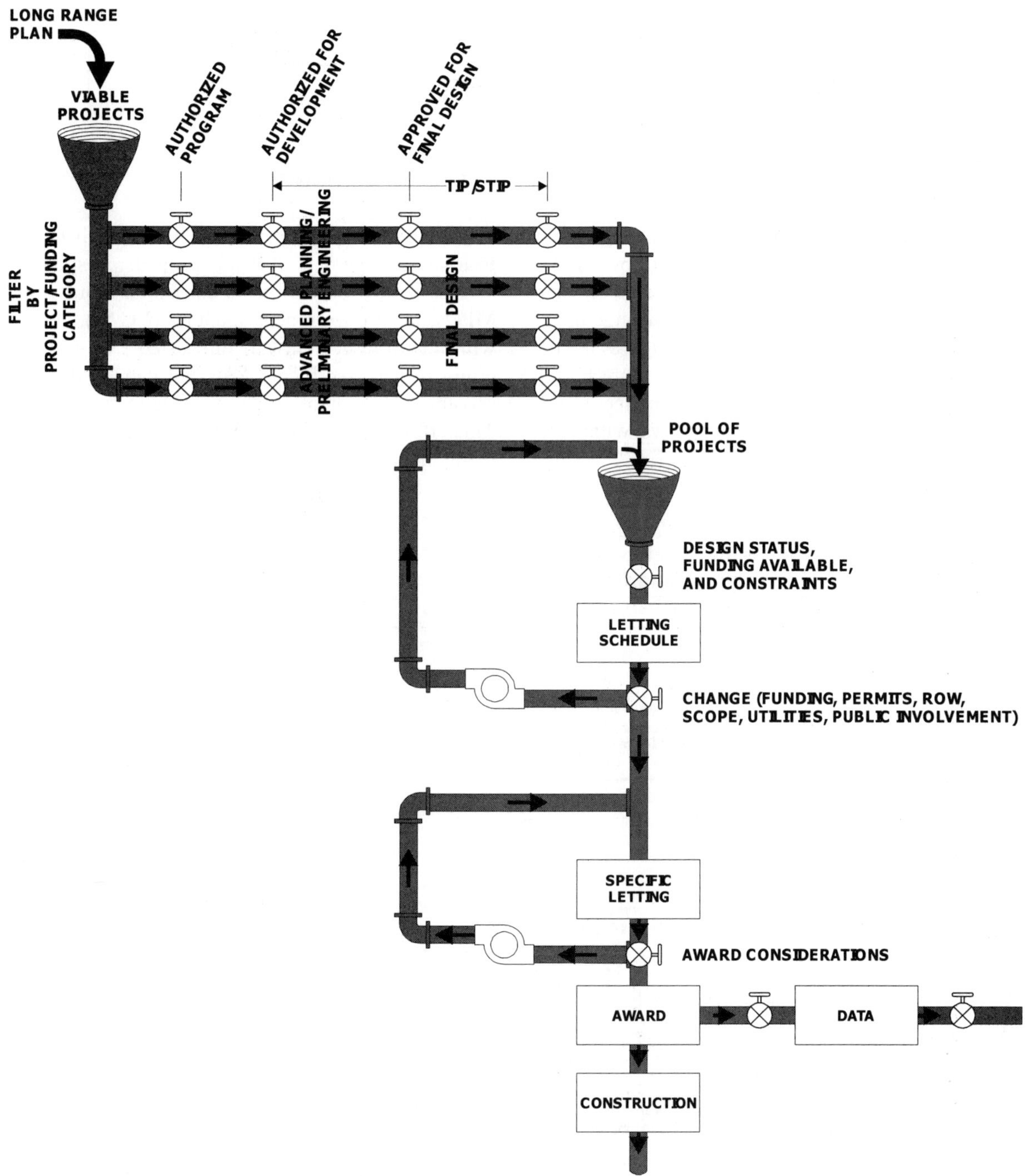

FIGURE 4 Illustration of a pipeline project.

picture that best represents components of SHA statewide highway letting program management was developed (see Figure 3). Key information required for these components is identified, as well as the interaction between components and different agencies or groups involved in letting program management.

Chapters three and four will describe in detail the specific aspects of developing and managing the letting program, as characterized in Figure 3 and schematically illustrated in Figure 4. References will be made to SHA practice as each component of statewide highway letting program management is discussed.

CHAPTER THREE

DEVELOPMENT OF LETTING PROGRAMS

INTRODUCTION

Before describing practices used by SHAs to manage their letting programs, it is necessary to understand how SHAs develop the letting program. This chapter focuses on that part of statewide highway letting program management. Figure 5 represents this part extracted from Figure 3. The components and information shown in Figure 5 are based on an analysis of responses to questions from Sections I and II of the survey questionnaire, interviews with five SHAs, and references from the literature.

BASIS FOR LETTING PROGRAM

Planning is the stage in the PDP that establishes the multiyear, long-range plan of viable projects, as shown in Figure 5. This plan is based on the combined inputs from SHAs, MPOs, local government agencies (LGAs), and the public with regard to current and forecasted long-range transportation needs. A critical needs list can be developed with possible solutions or viable projects to satisfy these needs. This effort is guided by state agency procedures and federal requirements. The multiyear program is formalized as a

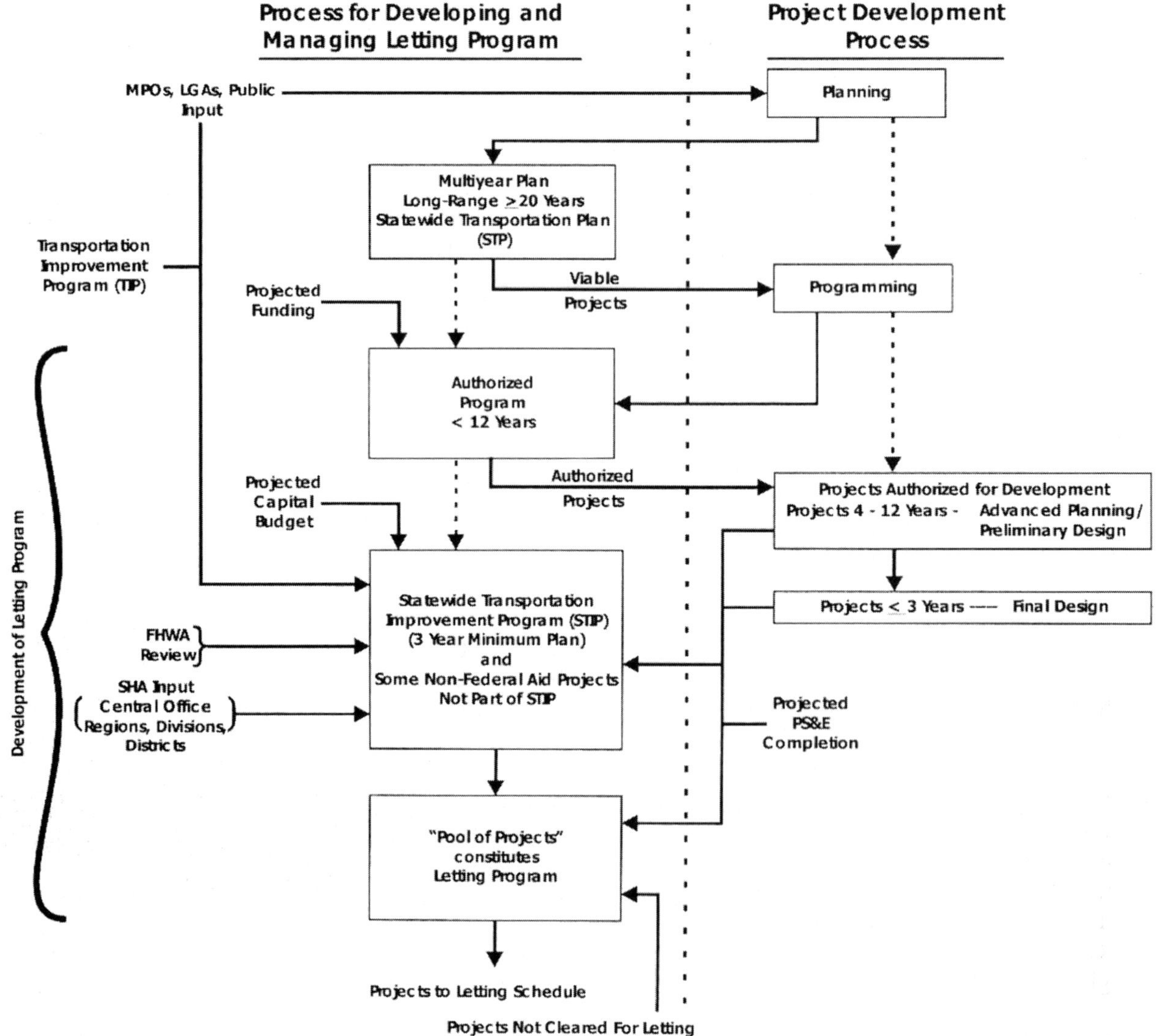

FIGURE 5 Development of a letting program.

Statewide Transportation Plan (STP) with a time frame of at least 20 years. Both the Georgia and Pennsylvania departments of transportation (DOTs) use a 25-year planning window. The STP is normally not fiscally constrained. This multi-year, long-range plan is not considered part of statewide letting program management and only serves as a source of input or a basis for developing the SHA letting program.

Viable projects are then prioritized and authorized for development consistent with the STP. This is accomplished through the programming phase of the PDP. Once a project is authorized, the advanced planning and preliminary design phase of the PDP can commence for that specific project. A group of authorized projects is considered the SHA's Authorized Program (see Figure 5). This program, according to questionnaire responses, is typically tied to a specific and briefer period such as 12 years or less. For the Texas DOT (TxDOT) this plan covers a 10-year period and is known as the Unified Transportation Program. The Washington State DOT (WSDOT) refers to its plan as the Capital Plan. It covers 10 years. The Georgia DOT has a 6-year Construction Work Program. The New York and Pennsylvania DOTs have 12-year transportation programs. The preparation of the authorized program initiates the development of the SHA letting program process for purposes of this synthesis in Figure 5 (see bracket on left side of the diagram). The next component in this process is the development of the STIP. The final component considered in the letting program process is the identification of the pool of projects, often referred to as the annual letting program. Each of these three components has a different planning horizon. The horizon for each component decreases starting with the 12-year authorized program. Once a project is authorized for development, a project identifier is frequently assigned to that project so it can be entered into a comprehensive database for tracking, such as the Project Support System used by the New York State DOT (NYSDOT) or the Design and Construction Information System (DCIS) used by the TxDOT. Some states, such as Connecticut, assign a planned letting date (month and year) for projects included in their authorized programs.

POOL OF PROJECTS CONSTITUTES THE LETTING PROGRAM

As shown in Figure 5, the focal point when developing a letting program is the STIP. Projects from the first 3 years of the authorized program usually form the basis for the STIP. Projects included in the STIP incorporate those projects from the TIP. The TIP is developed by MPOs with input from LGAs, the public, and, in many cases, the SHA districts, regions, or central offices.

The TIP and STIP are federal requirements. The STIP includes the TIP and additional projects outside of the metropolitan areas, sometimes called rural TIPs. Both the TIP and the STIP are financially constrained. The STIP may include funds for project phases other than construction. In any case, a project phase must be included in the STIP to be eligible for federal funds. Projects in the STIP considered ready for final design to support the construction phase are projected to have their PS&Es completed within the next 3 years. In some states, the STIP consists of only federal-aid projects. The STIP may include projects that are federally funded but performed by other agencies, that is, projects not in the SHA letting program. The STIP for the NYSDOT would include such projects. Other state-funded projects (100% state funds) may be considered, which are not directly incorporated into the STIP, as illustrated in Figure 5 (see also Figure 2).

Generally, the projects in the STIP are the first 3 years of the SHA's current authorized program. Some agencies, such as the Delaware DOT, use a time frame that is longer than 3 years. The Delaware DOT reported that

> A project is scheduled for funding in the 6 Year Capital Transportation Program (CTP). The first fiscal year of this program is the Department of Transportation Annual Capital Budget legislated in the State Bond Bill and signed into law by the Governor. The Bond Bill authorizes the use of the necessary funds to move forward on the scheduled projects.
>
> The CTP also acts as the State Transportation Improvement Plan (STIP), which is approved by the Federal Highway Administration (FHWA) and Federal Transit Administration (FTA) and authorizes the use of federal funds on the projects presented in the STIP.
>
> The actual obligation of federal funds and commitment of state funds are done on a per project basis.

Similarly, Georgia stated that "The yearly letting program is the first year of our 6-year Construction Work Program (CWP). The STIP is the first 3 years of the CWP."

Other SHAs, such as the TxDOT and Pennsylvania DOT (PennDOT), tend to focus only on 3 years—the minimum time required for the STIP. The TxDOT stated that, "Our 25 district offices are contacted in late spring and requested to provide projects to schedule in the next three fiscal years through an automated process." The automated process is TxDOT's DCIS, which tracks the status of projects as they move through the PDP. Input is controlled through a central division office but districts can input data as a project moves into the letting program pool of projects. The STIP must be updated at least every 2 years according to federal requirements. However, some SHAs update the STIP annually.

As Figure 5 indicates, a number of groups provide input for developing the STIP. MPOs develop the TIPs within their jurisdictions. Input from the public and other LGAs is considered in developing the TIP as well as SHA requests for specific projects at all levels including central office, regions, and districts. Once approved by the MPOs, the TIP is then incorporated into the STIP.

There are a number of general factors that influence the priority of projects in the STIP; however, the condition of the existing facility and safety issues related to the facility are perhaps the most influential driving factors that would give a resurfacing, restoration, rehabilitation, and reconstruction project a high priority. Furthermore, average daily traffic volumes, both current and projected, are considered when work to a specific existing facility is prioritized. Consistency with long-range plans and cost-effectiveness are also important factors, especially for new construction. In addition, budgetary considerations are carefully considered in the prioritization process, because the STIP is fiscally constrained. Finally, input from SHA regions, divisions, or districts can influence the priority of projects in the STIP.

Because the STIP is a dynamic program, projects that are closest to design completion will have the highest priority with regard to moving the project into the letting program pool of projects. Therefore, as shown on Figure 5, a projected PS&E completion date for the project, as determined by the status of final design within the PDP, is needed for all high-priority projects. The status of final PS&E approval is monitored closely by the SHAs. The Georgia DOT illustrated the linkage between the STIP and the letting program as follows:

> Every October, the department's top management begins to develop the annual program for the next fiscal year, which starts the following June. The yearly transportation program is developed by the department's top management from projects contained in the 6-year Construction Work Program based on objective areawide estimates of transportation needs, design capability, district construction workload, and congressional district fund balancing as required by state law. In urbanized areas, the MPO sets the priorities for their area, based on expected funding and project status information provided by the GDOT [Georgia DOT]. This annual element is approved by [the] GDOT Board as part of the STIP. For projects funded for construction in the first year of the STIP, the Preconstruction Division has monthly project review meetings to ensure that projects will be ready for contract let in the assigned month.

In general, an SHA will prioritize projects in the STIP and match project cost with budgets allocated during the upcoming 3 fiscal years. A federal-aid project must be included in the STIP. The FHWA will review the STIP to ensure that the projects in the STIP are eligible for federal funds. A projection of when final design of construction projects will be completed is necessary to compile the projects in the STIP. The STIP is approved and forms the basis for establishing the pool of projects. Interviews with respondents reveal that the first year of the STIP is most closely related to the pool of projects. Many SHAs consider this pool of projects as an annual letting program.

According to the questionnaire responses, there are a number of perspectives on how projects are included in the letting program. Several factors considered are the first year of the STIP, potential for project delivery (i.e., when PS&Es will be complete), status of ROW acquisition and utility adjustments, and environmental clearances. Furthermore, program funds allocated for the current fiscal year for projects, federal–state funding split, and congruence with department and district, region, and division goals are key drivers that move a project into the SHA letting program. The Connecticut DOT identified the following three factors that influence what projects are selected for its letting program:

- Departmental goals that, in part, address identified capacity and safety needs and a target number of bridges to be rehabilitated and lane-miles resurfaced;
- Order in which designs are completed for projects that can be funded through reasonable estimates of available resources; and
- Use of various reports to monitor design progress and construction funding requirements.

During the interview process for this synthesis report, the Connecticut DOT confirmed that the letting program is directly related to the first year of the STIP. Thus, projects are selected for initiation in cooperation with regional planning organizations, local officials, and the public. As previously discussed, this input is reflected in those projects included in the metropolitan TIP. Again, the TIP is incorporated into the STIP.

The Idaho DOT stated that a "balanced process based on readiness (extent of design completion) and funding available" is used to move projects into the letting program. Delaware employs a pipeline process that allows any person the opportunity to request a transportation improvement throughout the state. The project is assigned to a pool in which the project is prioritized on a statewide basis along with similar projects. The highest ranked projects are selected for the letting program. The WSDOT "uses a prioritization process based on getting the greatest change in program performance for the Highway System Plan objective per dollar spent." New York's approach to how projects are included in the state letting program from a broader perspective is described as follows:

> Within the program and fiscal parameters established by executive management, the Director of each of the Department's 11 Regional Offices selects projects for his/her region based upon established transportation goals, emphasis areas, and Commissioner initiatives,
> - an annual capital program target for lettings plus orders-on-contract,
> - a federal–state funding split, and
> - a breakdown of federal funds by type and amount.

The New Jersey DOT described the approach it follows to match funds to available projects:

> First, a pool of projects that have a high probability of going to construction is promulgated by the bureau of capital program development. Next, these projects are divided into a number of project lists according to funding category. The projects within funding categories are then prioritized based on each project's benefit to the state transportation system [known as the capital investment strategy (CIS)]. The annual program of projects that are to go under construction that particular fiscal year will number as many as there is available funding within each funding category.

Finally, the Maryland DOT described its letting program as a yearly endeavor:

> In the Spring of every year all projects in the Consolidated Transportation Program (CTP) have their cost estimates updated. These estimates include estimated costs for Project Planning (PP), Preliminary Engineering (PE), Right-of-Way acquisition (RW) and Construction phases of a proposed project. Based upon these updated costs the cash flows for those project phases that are funded are updated or revised. For those project phases that are not funded these estimates are used to establish our unfunded project need levels. From this information a Draft CTP is developed. Based upon available funds it is determined which projects to propose additional phases of project development to be funded and which projects to propose being added to the program. In the fall of every year the Department takes the Draft CTP "on tour" to the 23 counties and the city of Baltimore to discuss the Draft CTP and what priorities each jurisdiction has. After all tour meetings the Final CTP is put together. This Final CTP is submitted to the legislature in mid-January every year for approval.

As SHAs develop their letting programs, various tactics are considered to determine the number of projects included in the letting program. It is important to ensure that there will be sufficient projects with completed PS&E to use all available funds. One tactic often identified by SHAs as a means of achieving this goal is to ensure that there are projects available with their "plans on the shelf." Designs for these projects are essentially complete or 100% complete. Such projects can be incorporated into the letting program if and when funds become available. For example, the WSDOT develops a list of "Advance Engineered" projects that can be substituted for any project delayed during the project development phase or in the event of additional spending as a result of low bids or increased spending authority from the state legislature.

Some states also referred to this tactic as overprogramming. The percentage of projects that are overprogrammed ranged from a low of 5% to a high of 20%. These projects are not necessarily included in the STIP, because the STIP is fiscally constrained. However, projects can be added to the STIP by an amendment if other projects will not be sufficiently developed to meet the proposed target letting date. SHAs that specifically referred to this tactic included Georgia, Louisiana, Minnesota, and Texas. For example, Georgia stated that, "Our annual programs are developed based on a budget that equals 100% of our appropriation from FHWA. Usually, our obligation authority equals between 80 to 90% of our appropriation. Therefore, we are usually over programmed by about 10%." The Minnesota DOT indicated that its program is overprogrammed by 6% to 9%. In the interview with the TxDOT, staff there reported that the letting capacity for each district is overprogrammed by 20%. However, when the district reaches 80% of its letting capacity, it submits remaining projects for review, or projects on the shelf, in hopes of having additional funds to carry some of those projects forward to letting.

Most other SHAs stressed that there were always sufficient projects to meet current capital budgets. Therefore, from these agencies' perspectives overprogramming was not necessary. These SHAs noted that they can always complete sufficient PS&Es for projects in their letting program to use available funds. For example, the California DOT (Caltrans) stated, "There is never a shortage of projects. Occasionally, there is a shortage of funds, in which case we wait until the next state budget cycle." The New Hampshire DOT indicated that having projects close to completion and overprogramming can work together: "We first check the Status of Funds report from the FHWA. We then use 'on shelf list'—[a] list of projects that are readily available if money appears. We have typically overprogrammed our project list so using funds is not a problem."

PennDOT attempted to develop a manual that was related to the interaction between the PDP and the transportation program. A particular focus was on processes that involve funds management. A draft document, *Letting Capacity Enterprise-Wide Gap Closure* (2002), was developed that captures these processes.

PennDOT recognized the need to make efficient use of limited resources by improving the overall project and funds management process, as well as clarifying associated policies and procedures to minimize the risk of shutting off new lettings or being unable to pay bills. Gap closure teams consisting of staff representing highway operations, fiscal services, and planning and programming at the district and central office levels were assembled. Their assignment was to develop an integrated, consistent letting capacity process aligned with department goals and linked to the overall transportation program (*Letting Capacity* . . . 2002). The 10 functions proposed in the manual were (1) Pennsylvania's transportation program, (2) obligation plan, (3) multimodal project management systems (MPMS) cash flow, (4) district project management checkbook, (5) commonwealth budget, (6) district project monitoring, (7) accrued unbilled costs, (8) capital budget, (9) let schedule, and (10) data management. Each function was described according to actions, by whom, when, rules, measures of success, key players, and reports. An example of one function is shown in Table 3.

TABLE 3
PENNSYLVANIA'S TRANPORTATION PROGRAM (from *Letting Capacity Enterprise-Wide Gap Closure* 2002)

Action	By Whom	When
Determine available federal and state capital funding for next Transportation Program update; work with Planning Partners to develop Program Financial Guidance	Program Center Fiscal Management	July 31, odd numbered years
Issue overall Program Update Guidance to Planning Partners and Districts	Program Center	August 31, odd numbered years
Public input via State Transportation Commission/Planning Partner public hearings	State Transportation Commission, MPOs, LDDs, Independent Counties	November 15, odd numbered years
Develop preliminary draft regional TIPs	MPOs, LDDs, Independent Counties, Districts, Program Center	February 1, even numbered years
Provide PennDOT comments on preliminary draft TIPs	Program Center	March 31, even numbered years
Air quality conformity analyses, public comment periods, other federal transportation programming requirements (environmental justice)	MPOs, LDDs, Independent Counties	June 30, even numbered years
Adopt TIPs	MPOs, LDDs, Independent Counties	July 15, even numbered years
Approve 12-year program	State Transportation Commission	August 15, even numbered years
Submit recommended program (STIP) to FHWA, FTA, and EPA	Program Center	August 15, even numbered years
Approve STIP and air quality analysis	FHWA, FTA, and EPA	September 30, even numbered years
Begin to implement projects	Districts	October 1, even numbered years

Rules

TIPs must be fiscally constrained by year based on the levels of funding and the types of funding. Federal funding for the TIPs and the STIP is based on the level of federal funding authorized in the most recent federal surface transportation act. The level of state funding is based on state revenue projections and the resulting budget projections developed by the Bureau of Fiscal Management.

Project cost estimates and schedules in MPMS must be current and accurate for each project phase.

Program adjustments must be approved by PMC and the MPO/LDD/independent county in accordance with established operating rules. These operating rules are adopted locally by the MPO, LDD, or independent county. The Department is an active participant in that process and a member of the MPO, LDD, and independent county transportation program governing body.

Program adjustments must be consistent with the amount and types of funding available. When Districts and planning regions have difficulty in balancing funding types within TIPs, the Districts may solicit the assistance of the Program Center to attempt to make the adjustments with other regions and Districts on a statewide basis.

Measures of Success

Program development guidance is developed on time
STIP approved by September 30 of even numbered years
Minimal need for TIP changes in October due to project slippage from the TIP approved in June/July to that which takes effect on October 1

Key Players

Program Center
District Planning & Programming Manager
Assistant District Engineer—Design
District Engineer/Administrator
District Portfolio Manager
Fiscal Management

Reports

MPMS TIP 200 report—single line listing of project phases and costs by year for the TIP selected.
MPMS TIP 222 report—a multi-line listing of project phases and costs by year for the TIP which is selected
Customized reports
Appropriation 185 letting capacity cash flow analysis

Notes: MPO = metropolitan planning organization; LDD = local development district; TIP = Transportation Improvement Program; STIP = Statewide Transportation Improvement Program; EPA = U.S. Environmental Protection Agency; MPMS = multimodal project management systems.

SUMMARY

The development of the letting program is complex. Many agencies and the general public are involved in the letting program process and provide input as to which projects are included in the STIP and their priority. Key factors that influence the priority of projects in the STIP include safety, level of traffic (average daily traffic), and condition of the existing facility. Development of the STIP is closely tied to the PDP with respect to the status of design completion. The first year of the STIP is the primary determinant of which projects are included in the pool of projects. However, there are many different ways in which SHAs select specific projects for inclusion in their letting program. There appear to be some dominant factors that influence this decision, such as delivery status of the project, project priority established by SHA districts and divisions with input from the FHWA, and availability of funding.

CHAPTER FOUR

MANAGEMENT OF LETTING PROGRAMS

INTRODUCTION

Once the letting program pool of projects is developed, the management of the letting program begins. The components and information required to manage the statewide highway letting program are shown in Figure 6, which provides a detailed representation of the process followed to manage the letting program, as extracted from Figure 3 in chapter two. The components and information shown in this figure are based on the analysis of responses to questions from Sections I through V of the questionnaire, interviews with five SHAs, and references from the literature.

LETTING PROGRAM AS THE BASIS FOR THE LETTING SCHEDULE

A major action in letting program management is to establish the letting schedule. Typical information on the letting

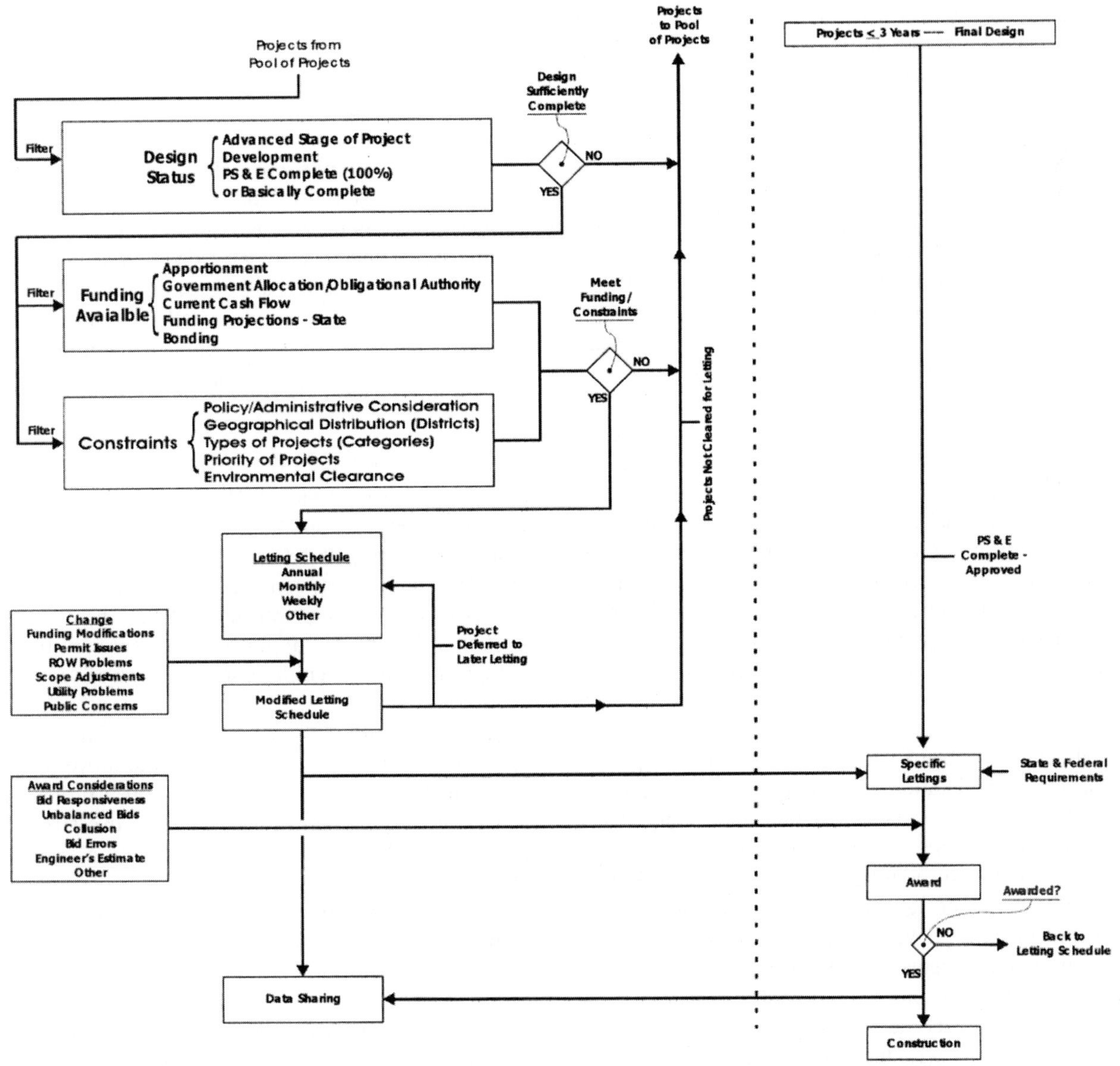

FIGURE 6 Management of a letting program.

TABLE 4
FACTORS CONSIDERED FOR SCHEDULING PROJECTS FOR LETTING

Category	Factor
Design completion	Design nearing completion and/or production ready (approved for letting); also includes status of right-of-way acquisition, utility adjustments, and environmental clearance
Funding	Availability
Constraints	Construction needs—sequencing and timing of project, seasonal considerations, pricing issues Balanced lettings—number of projects, type of projects Addenda required—time needed to process Priority—consistent with state annual program Location—in which region or district project is located Safety issues—concern for accidents Project complexity—type, cost, difficulty, duration Contractor request—volume, pricing

schedule could include the location of the project, responsibility for the project (e.g., district), brief description of the project, target letting date, and cost range (e.g., see PennDOT website in Appendix B). The letting schedule may simply identify dates (month and day) when project lettings will be conducted, together with a brief description of the projects to be let on those dates. Most often, more specific information concerning projects to be let is provided in supplemental documents, such as letting information for progressing projects and the notice to contractors (e.g., see Georgia DOT website in Appendix B). These letting schedules cover different time frames, such as Arkansas's 1-year moving schedule or PennDOT's 6-month schedule, which is updated at the end of each month. Most SHAs post letting information on their websites.

As shown in Figure 6, there are three main categories or drivers that influence when a project is placed on the letting schedule: (1) status of design completion, (2) funding available, and (3) constraints. These three categories cover a number of different issues considered when selecting projects for the letting schedule and confirming the target letting date assigned to that project. They act as filters in regard to placing a project onto the letting schedule. Table 4 shows key factors considered when placing projects onto the letting schedule.

As reflected in Table 4, the status of design completion is the initial consideration. According to the interviews with SHAs, this may be the most important filter that a project must pass through to move onto the letting schedule. Design completion is reflected in Figure 6 as flowing through the filter of design status. Design completion covers all facets of PS&E development, including ROW acquisition, utility adjustments, and environmental clearance. Extensive advanced planning is required to ensure that the letting schedule reflects the best target dates for each project letting, consistent with the status of design completion. If the design is not sufficiently complete, the project will remain in the pool of projects for future consideration (Figure 6).

Once the project is deemed ready, there are two other filters it must pass through before being placed on the letting schedule (see Figure 6): funding availability and constraints. Funding depends on the source, current cash flow, and future projections. For federal-aid projects, funding requirements must follow the guidelines set by the FHWA ("Financing Federal-Aid Highways" 1999). In addition to having funds allocated by the FHWA, the state's portion of the project funds must also be available for the project. This determination is based on both current cash flow (i.e., expenditures in the current fiscal year) and projections for the remaining years in which the project will be under construction.

As shown in Figure 6 and identified in Table 4, certain constraints also act as a filter. These constraints could be related to policy and administrative issues, geographical distribution, project types, and priority. SHAs identified specific factors they consider when determining if a project is ready to move onto the letting schedule (see Table 4). Other factors that might be considered are bidding volume, availability of contractors, and time of year (season).

The availability of sufficient funds and satisfying the applicable constraints allow a project to move on to the letting schedule. As shown in Figure 6, if a project does not satisfy the requirements posed by these two filters then the project would not be scheduled for letting and would remain in the pool of projects. Most SHAs consider a combination of factors from the three categories of status of design completion, funding available, and other constraints. The following are some specific examples:

- Arizona—projects are scheduled based on an extensive planning process, geographical (seasonal) issues, as well as funding availability.

- Arkansas—projects are assigned to lettings based on location and need, availability of funds, project priority, design status, safety considerations, status of permits and agreements, and ROW availability.
- Kansas—once the program year is set, the letting month is determined by considering project development time constraints and construction sequencing or timing. The number of construction seasons required for the project is minimized whenever possible.
- Louisiana—when design is complete and funds are available.
- Maine—a combination of departmental priorities among projects that are ready to proceed and funding availability.
- Minnesota—we try to "balance" lettings over the year so that all projects are not let in the spring.

As indicated from these examples, certain factors have greater importance than others when placing projects on letting schedules. The NYSDOT provided this comprehensive explanation:

> Central Office tries to let projects on the letting dates requested by the Regions. These are based upon scheduled PS&E completion (which allows design work to begin on the next project on schedule) and scheduled completion of the construction work. For example, projects that can be completed in a single season need to be let in a particular time frame or they may be delayed until the following year to avoid becoming a two-season job.

In addition to the letting date requested by the Regional Office, Central Office considers:

- timely completion of PS&E so that bid documents can be prepared and made available when needed;
- funding availability (annual state funding is frequently made available in half or quarter year increments, delays in state budget approval);
- number of projects in a letting (staff limitations keep this to 25–30 per letting);
- is additional time needed for an amendment(s);
- contractor wishes (bids are sometimes delayed to accommodate contractor's request for delay if reasonable, for example, letting conflicts with an industry conference, several large projects are scheduled to be let in too close a proximity);
- uncertainty in another part of the statewide letting program (e.g., a project scheduled by one Region that might not make the end of year cutoff may cause us to schedule an equivalent amount of supplemental letting to assure annual statewide letting total is reached);
- uncertainty over the amount of orders-on-contract (OOC). The Department is given a combined letting/OOC limit. If OOCs are more or less than expected, the amount of lettings must be adjusted.

Finally, as shown in Table 5, states have different frequencies for lettings. The "Other" category in Table 5 would include approaches such as having 20 lettings per year.

TABLE 5
FREQUENCY OF PROJECT LETTINGS

Frequency	No. of States
Monthly	11
Bi-weekly	4
Weekly	8
Other	5
Total	28

CHANGE IN LETTING SCHEDULES

As discussed in the previous section, SHAs have letting schedules that are typically based on a calendar time frame such as a year, 9 months, or some other duration. This letting schedule is often revolving, with new projects being added while other projects are removed. Projects that are removed from the letting schedule most often have been let and awarded for construction. However, some projects are removed from the letting schedule owing to circumstances or events that would prevent the project from meeting its specified letting date. The letting schedule is often modified to reflect the impact of these circumstances or events. These modifications can occur as frequently as monthly. Decision makers must manage the impact that change has on project letting dates, especially when the project cannot be rescheduled within a short period for letting.

SHAs were asked to identify major factors that bring about change in the letting schedule. Factors identified by respondents were summarized and grouped into 11 different areas. These 11 areas are shown in Table 6, in addition to the frequency with which each area was cited as a factor that changed the letting schedule.

For each factor area shown in Table 6, the respondents identified the causes of change. Actions taken to mitigate the impact of change as related to projects on the letting schedule were also specified. Table 7 highlights the major causes of change for the four highest cited factor areas; that is, funding or cost, environmental clearance, ROW, and project scope. Typical actions taken by SHAs to mitigate the impact of change are also shown. Inspection of Table 7 shows that many of the causes cited are shared among the other factor areas.

The causes of change appear to be associated with two categories of issues: those in which the SHA may have some control over the cause of change and those that are perhaps beyond SHA control.

Issues where the SHA may have some control over the cause of change in the letting schedule are related to managing design modifications that affect funds or require additional work. Another issue is the impact resulting from not meeting deadlines or being too optimistic about achieving

TABLE 6
TYPICAL FACTORS THAT CREATE CHANGE IN LETTING PROGRAM

Factor Area	Times Cited	Typical Factors Cited
Funding and/or Cost	21	Budget shortfall or lack of funds, cash flow uncertainty, changes in state budgets, cost creep, changes in matching funds
Environmental Clearance	14	Obtaining timely permits because of readiness, clearance, new regulations
Right-of-Way	12	Difficulty in acquiring ROW, delays, readiness
Project Scope	8	Changes in specifications, program changes in TIP/STIP, changes directed by executive office, changes in demographics or traffic patterns in project area, changing site conditions
Utilities	6	Coordination on federal projects, relocation not complete, clearance not acquired
Design Completion	6	Delays in design progress with plans not ready to go, conflicts with ROW and utilities, addenda during bidding, workforce shortages, work overload
Schedule Constraints	6	Slippage as a result of not meeting deadlines, late information, lack of timely and clear PS&Es
Project Priority	5	Shift in priorities from legislature, emergency projects push others out of schedule
Interagency Coordination	4	Agreements not signed, cost too high, impact of third party involvement
Plan Accuracy	3	Mistakes, clarifications, processing errors
Projects Status	3	Project not required; clearing non-lets

Notes: ROW = right-of-way; TIP = Transportation Improvement Program; STIP = Statewide Transportation Improvement Program; PS&E = plans, specifications, and estimates.

TABLE 7
TOP FOUR FACTORS CAUSING CHANGE IN LETTING PROGRAM

Factor Area	Main Causes	Typical Actions
Funding and/or Cost	Lack of resources to fund projects; overruns of other projects; trying to let too many projects; changes in funding type; lagging schedules shifted funds to other projects; additions/reductions in program allocations; increased spending in congestion relief through transfer of funds; more complete information; poor initial estimates; government bodies such as state legislatures modify funding levels; design changes increasing cost; costs associated with schedule delays; changes dictated by field actions	Reschedule projects (move to next fiscal year or bring projects into current fiscal year); overprogram to use all available funds; change funding source or obtain additional funds; develop better early estimates of project cost; better manage cash flow to balance dollars; modify project scope by adjusting limits; minimize changes; select from committed projects or projects on reserve; tapering of federal funds, that is, use federal funds first; monitor large projects that may slip; hold lettings to minimum target but maintain contingency
Environmental Clearance	Delay in obtaining federal approval of environmental documents; lack of staff/high workload; process delays or too late in process; lack of training; simply not meeting deadlines; process/coordination takes longer than anticipated; unforeseen problems related to endangered or threatened species, archeology, Native American, historical features	Move effort to earlier in project development process; improve process of obtaining permits; closer coordination with agencies may be through meetings; hire additional staff; place project on hold (move back to pool of projects)
Right-of-Way	Delay by relocatees; condemnations and legal issues obtaining property; not 100% ready to certify by FHWA representatives; late design changes; schedule delays do not leave adequate time to purchase land; increase cost of ROW; late land surveys; too optimistic in timing of acquiring ROW; lack of permits; lack of trained staff	Reschedule projects; shelve projects; move effort to earlier in project development process; conditional certifications of ROW; restrictive clause to limit contractor access; improve coordination schedule between design and management; closer coordination with agencies; improve project management process and provide new tools
Project Scope	Demographic/traffic pattern changes result in adjustments in design requirements; community input or public process; unforeseen problems such as changed site conditions; changes in specific designs such as pavement or traffic control; design team decision to modify scope; modification to design standards	Reschedule projects; process changes in timely manner; limit late scope changes in final design

Notes: ROW = right-of-way.

certain project milestones that are critical to a project's meeting the letting date. Some issues in which the SHA may not have substantial control over the cause of change could include directives from state legislatures (more or less funds available), delays from lack of federal approvals of design or environmental-related documents, delays in acquiring ROW owing to legal requirements, and changing demographics or traffic patterns.

Actions taken in the four critical factor areas shown in Table 7 can be generally categorized as preventive or reactive. A preventive action would likely be a response to those causes of change that may be within the control of the SHA decision maker who is involved with managing the letting schedule. For example, if additional funds become available for project work because project bids are lower than the engineer's estimate, then overprogramming may be one action that will ensure that these funds can be used. One example of overprogramming is having projects "on the shelf," which can quickly be moved onto the letting schedule and assigned a specific letting date. Conversely, a reactive action would more likely respond to causes where the SHA decision maker has little or no control over the situation, such as legal issues related to ROW that might delay a letting. The action taken in such a situation would be to reschedule the project. This action may move the project to a later letting date within the fiscal year or to another fiscal year, or it may lead to the project's removal from the letting schedule completely. Table 8 provides similar information on causes and actions for the remaining seven factor areas.

The causes and actions taken as listed in Tables 7 and 8 lead to several general impacts to the overall letting program. The primary impact is reshuffling of projects to ensure that allocated funds are used. Thus, SHAs will move projects into and out of their letting programs. Also, specific letting dates are modified, with some project lettings delayed while others are maintained. In a specific case, the NYSDOT uses a tapering concept (federal funds first) with federal-aid projects to ensure that the more costly federal-aid projects are let on schedule. SHAs also have to exercise caution in the number of projects included in any one letting, because a higher number of projects let at the same time can result in higher bid prices.

The challenge appears to be in maintaining consistency in letting projects. Meeting that challenge requires monitoring of letting goals. The New Mexico Department of Transportation Department (NMDOT) uses as a tool a document that helps everyone in the agency assess how well the department is delivering services and products to its customers ("Compass 2nd Quarter" 2002). One section of NMDOT's tool provides performance data related to the STIP. Graphs and charts are used to show performance in the following areas:

- Number of programmed projects let,
- Dollar amount of programmed projects let,
- Actual bids versus programmed amounts,
- Bid amount within 10% of engineer's estimate,
- Actual cost versus low bid amount, and
- Programmed cost versus actual cost.

The data contained in this tool are updated every quarter. Appendix D provides examples of these graphs and charts. NMDOT management can monitor these graphs for improvement or to identify problems that require corrective action, to meet specified performance targets.

Change may result in movement of a project's letting date within the letting schedule, and it may return the project to the pool of projects. Most agencies revise their letting schedules frequently. As previously stated, some SHAs use a moving 12-month schedule, whereas others use a 6-month schedule that is updated monthly.

Interviews with SHAs indicated that regular meetings are a typical technique used to manage the letting program and schedules—monthly by the NYSDOT and weekly by the

TABLE 8
OTHER MAJOR FACTORS CAUSING CHANGE IN LETTING PROGRAM

Factor Area	Main Causes	Typical Actions
Utilities	Lack of obtaining permits in timely manner; lack of agreements for utility relocation including funding; advanced utility relocations not complete; coordination slow	Reschedule projects; shelve project
Design Completion	Schedule is too optimistic—cost, complexity; addenda due to late design changes/cost increases	Delay project; find ways to maintain schedule; use consultants; overtime for staff processing PS&Es; cease all amendments
Schedule Constraints	Design team not meeting schedule or deadlines; inadequate resources; inadequate information provided by field not providing timely input; unrealistic schedule; inability to meet advertising requirements	Adjust letting schedule; start project earlier; provide more realistic deadlines by using scheduling software; dedicate more resources or work overtime; use special lettings; minimize changes that impact schedule
Project Priority	Competing priorities; changes needed; accommodation of emergency project (pushes project to lower priority)	Use consultants; adjust letting schedule
Interagency Coordination	Coordination with other agencies slow process; last minute changes to obtain required authorizations or third party agreements; railroad coordination delays	Increase early involvement with agencies
Plan Accuracy	Design flaws; unexpected subsurface conditions	Process addenda; redesign
Projects Status	Traffic patterns change and/or other projects in vicinity corrected problem; small projects swallowed up by other projects	Overprogram, that is, have projects on the shelf

Notes: PS&E = plans, specifications, and estimates.

TABLE 9
TYPICAL STATE BIDDING PROCESSES

State	Bidding Process
California	Projects are advertised to the public and plans and specifications are available for purchase for 4 weeks prior to bid opening (shorter for emergencies & longer for large complex projects). Bids are opened on Tuesdays and Wednesdays in Northern California and on Thursdays in Southern California (so that local contractors don't have to travel several hundred miles north to Sacramento). Bids are publicly opened and read; then they are mathematically verified. The proposal books are then checked for completeness, contractor's licenses are checked for proper category and currentness. The public and other Department personnel are then notified of the results.
Delaware	In basic terms, without a pre-bid meeting, a project is processed as follows: advertised for two consecutive weeks; 30 days between ad date and bid opening; documents available to bidders for a minimum of three weeks; questions are entertained during this time period; questions can result in addenda depending on the nature; bids are opened on Tuesdays; bid documents reviewed for compliance and completeness; bid tabs are prepared and sent to the originating section for recommendation on award; if recommended for award, finance prepares the obligatory funds; agreement and bonds are sent to successful bidder; the successful bidder must return the agreement and bonds within 20 calendar days; after return, the documents are executed by the Department.
Maryland	*Contract solicitations are formally advertised a minimum of 20 days prior to bid opening. A Notice to Contractors is placed in the *Maryland Contract Weekly*, *The Daily Record*, and *The Afro-American*. Some projects are preadvertised, making advance copies of nonfinalized plans available on CD-ROM for review and comment (but not for bidding purposes). *Pre-Bid conferences are not required and when they are held attendance cannot be made mandatory. Procurements in excess of $100,000 require that a written record of the conference must be sent to all prospective bidders. *Addenda must be issued to give bidders sufficient time to prepare and receipt of all addenda must be received prior to bid opening. *Bids for $100,000 or more must be accompanied with a bid security of 5% of the total bid amount. This can be in the form of a Bid Bond, Certified Check, Cashier's Check, or Cash. *All successful bidders must be registered with the Maryland State Department of Assessments and Taxation and possess a valid State of Maryland Construction License. *All bids are verified through AASHTO's Trns•port System and the project is awarded approximately 30 days after the bid opening. During this 30 day period, an Experience & Equipment form and the Affirmative Action Plan, if necessary, is submitted by the low bidder and approved by the State. If the bid is 10% over or 15% under, the design engineer must prepare a Justification letter for the procurement officer's signature. *After award, legal binding Contract Documents are issued to the low bidder for execution of the affidavit and of the Payment & Performance Bonds for 100% of the contract price. After obtaining all low bid and SHA signatures, Notice to Proceed is given for commencement of work.
Texas	Prospective bidders request bid documents from the department. Bid documents are sent to bidders meeting the pre-qualification requirements. Bid proposals are sent out approximately 21 days prior to the bid opening date. Electronic proposal plan and bid item data are posted to the Internet at this time. Bidder's list data, etc., are also uploaded. Project addenda information is mailed to prospective bidders and posted on the Internet on Tuesday, the week before the bid opening. Bids are received and opened on the bid opening date. The bids are tabulated and the results are posted on the Internet. Subsequent to the bid opening, the bids are evaluated and recommendations for award or rejection are prepared and reviewed as necessary.

Arkansas DOT. Various reports are used to provide key information about the status of projects as they progress through the letting process. Computer-based tracking systems provide much of the information needed, such as the Project Support System used by the NYSDOT or the DCIS used by the TxDOT. Other tools are used to manage money (Financial Management Information System) or schedule (Primavera). Databases are often used for letting schedule management. The Connecticut DOT described funds management this way:

> The Bureau of Engineering and Highway Operations maintain a Capital Construction Program Schedule for projects without federal participation and a Working Schedule of Federal-Aid Obligations for project phases with federal participation. These two lists are used as agendas at a monthly Project Status Review meeting to update project cost estimates and schedules. The resulting constantly changing project lists are the basis for the Department's programming of available funds. Simultaneously, the Department maintains state bond listings that identify the bond authorization to be used to fund the non-federal share of upcoming projects. This process allows frequent input from the various offices involved that may impact the letting schedules.

IMPACT OF AWARD CONSIDERATIONS ON SPECIFIC LETTINGS

Although the steps covered in the bidding process are similar among SHAs, there are variations that are apparent and, hence, the bidding process is customized to reflect SHA practice. Some examples of this process are shown in Table 9.

Most bid lettings lead to successful award of contracts. However, there is certain information that SHAs examine to ensure that the award is properly made to the lowest responsive and responsible bidder. Approximately 70% of the survey respondents stated that their agencies may not always award the contract to the low bidder. Otherwise, if the project is awarded, it is awarded to the lowest bidder. Table 10 provides typical reasons why the lowest bidder would not be awarded the bid.

Several issues identified in Table 10 are addressed during bid analysis in a number of ways. Bid analyses may result

TABLE 10
TYPICAL REASONS FOR REJECTING BIDS

State	Reasons for Rejecting Bid
Ohio	Failure to submit an intact bid proposal; to submit a bid guaranty; to sign the bid; to include addendum; to include a supplemental questionnaire; to sign and notarize the supplemental questionnaire; to sign the bid bond; omission of unit price of lump sum; not pre-qualified; over pre-qualify dollar amount.
Maryland	Maryland will not award a contract to the Low Bidder if the firm does not submit a responsive and responsible invitation for bids package; if the firm does not submit, enclosed in the sealed bid, a responsive and responsible 5.00% bid security; if the vendor wants to withdraw and has adequate proof and documentation of its bid error; if the vendor is not formally registered with the Department of Assessment & Taxations; or whenever it is not in the best interest of the State. If and when these scenarios arise we may award the contract to the next lowest responsive and responsible bidder or re-advertise the project.
Nevada	If bidder's preference applies (state-funded contracts only) and the low bidder's bid is less than 5% of the next bidder's bid and the bidder does not qualify for the preference, we will award to the second bidder provided his bid is not unbalanced or over 7% of the engineer's estimate.
	If documents required in the bid, that is, bid bond, anti-collusion affidavit, or subcontractor listings are missing. If DBE goal or good faith effort is not attained. Protests on contractor licensing for prime contractor or subcontractors are received.
Texas	Unbalanced bidding, total bid price too high, inadequate bidder competition, and error in bid by contractor.

Notes: DBE = Disadvantaged Business Enterprise.

in rejection of the bid. For example, 78% of the respondents stated that their agencies have criteria for rejecting bids that overrun or underrun the engineer's estimate. Several states use a single percentage, such as 5%, 7%, or 10% over the engineer's estimate. Other states use ranges expressed by both under and over the engineer's estimate, such as 20% below and 10% above or 15% under or 10% over. If the project is outside the specified range, then justification for the award must be provided. For example, the Maryland DOT noted that

> Bids 10% over or 15% under the engineer's estimate require a written justification, which must be approved and signed by the procurement officer. If the bid is substantially above the 10% allowance, the design office may write a justification letter, requesting all bids be rejected. The letter, including a detailed explanation of the discrepancy, must be approved by the Assistant Attorney General's Office and the procurement officer. When a bid is rejected, notification is distributed to all bidders.

In some states, state law dictates the percentage and resulting actions; for example, in Ohio, "Pursuant to Ohio Revised Code 5525.10, ODOT [Ohio DOT] may not award a contract for a sum greater than 5% over the estimate." The Kansas DOT follows another approach, which is not directly tied to a percentage over or under the engineer's estimate:

> We do not automatically reject bids that are over the estimate. If a bid is over the estimate we handle the review on a case by case basis. There may be cases where the estimate was too low due to conditions that were not known at the time the estimate was prepared. There may be other factors. We look at the bids and if we decide it is to the state's advantage to award rather than reject and relet, we will do that. One bid does not cause that bid to be automatically rejected. Since we have the estimate, we can use that for comparison of the bid.

Another area of concern in bid analysis and award is unbalanced bids. Most agencies surveyed (26 of 28 states responding) conduct some level of analysis when bids appear to be unbalanced. Eight states use a computer program, such as the AASHTO Trns•port software, to support this analysis. The Maryland DOT reported that it

> . . . utilizes AASHTO's Trns•port System, DSS Module (Decision Support System). This software is used to create a graph and listing of the bids compared to the engineer's estimate and to each other. We can also utilize the system to create reports on past bidding history, market prices, price differential in geographical locations, etc. Maryland has a BAMS (Bid Analysis Management System) team, which is made up of individuals from several disciplines, such as, [the] Construction, Bridge, Highway, Design, and Information Technology divisions. The team is to meet on a regular basis to review the graph and listing of pertinent Maryland projects. In addition to the team's review, the administering design office also reviews the same materials. Recommendations are made and documented as needed.

The TxDOT takes a slightly different approach:

> We have developed a computer program that analyzes certain items in the bid to determine if unbalancing may have occurred. If the program indicates that unbalancing did occur, further analysis is done using [the computer program] Primavera. The schedule developed by the department to estimate the job is used with the contractor's unit bid prices. The net present value of the cash flows is calculated. If the cash flows reveal that the second bidder may be lower, we give the low bidder the chance to present their case using their construction schedule. If the cash flows still favor the second bidder, the bid is rejected.

Other states use procedures that require manual calculations when analyzing unbalanced bids. These procedures are typically documented in PDP manuals (New York and Pennsylvania) or referenced in FHWA publications (Pennsylvania):

New York—Draft Manual of Administration Procedures (MAP 7.1-5), Comparison and Evaluation of the Low Bid with the Department's Construction Cost Estimate.

Publication 408—"Specifications," Section 102.13 (website same as Design Manual Part 1); Also see Contract Administration Core Curriculum Participants Manual and Reference Guide 2000 at www.fhwa.dot.gov/infrastructure/progadmin/contracts/index.htm.

Collusion is another area of concern. Of the states responding to the questionnaire, one-half reported that they have a procedure to identify whether or not collusion might be present in the bidding process. States that had a procedure as well as those that did not reported that contractors were required to sign a statement or affidavit certifying under penalty of perjury that there was no collusion in the preparation of their bid. This is an FHWA requirement for federal-aid projects. A specific form is included in the bid package. Several states use the AASHTO Trns•port software to help analyze bids for collusion (Decision Support System) or the reports from the Bid Evaluation and Line Item Profile.

Each state was asked to describe how its agency determines if a bid is not complete and/or nonresponsive. Generally, SHA staff will review every bid submittal and will check for required signatures and documents included with the bid. Some specific items considered include the company seal on the signature page, attachments and inclusions of all addenda issued, proper bid security, and complete schedule of prices. Caltrans considers bids to be nonresponsive if they have one or more of the following problems:

- The unit price and the extended price for any particular bid item are left blank.
- The terms and conditions of the bid are modified or qualified by the bidder.
- An acceptable bidders security is not present with the bid.
- The bid itself is unsigned.

All other statements and certifications in the proposal can be made or corrected after bid opening, at the discretion of the department.

The Kansas DOT considers bids incomplete or nonresponsive if one of the required forms is not completed, signed, and/or notarized. These forms include the Proposal, Bid Bond, Certification of Noncollusion, Certification of Prequalified Financial Amount, Certification of Contractual Services with a Current Legislator's Firm, Declaration Limitations on Use of Federal Funds for Lobbying, Disadvantaged Business Enterprise Contract Goals, and if any of the Unit Prices are not included, or if a unit price is not changed correctly (in ink with the initials of the person making the change). The standard specification states that we (Kansas DOT) use the unit price as the intent of the contractor.

Many SHAs also referenced specifications that list those requirements for bid acceptance. These specifications can often be found on websites (e.g., Georgia; see Appendix B).

Additionally, SHAs were requested to explain their policies on errors in bids. In general, it was noted that staff and legal counsel review bid mistakes, as appropriate. Minor errors are often waived, but material errors may not be waived. Depending on the error, the bid may be declared irregular. If it is an error that can be fixed, the contractor may be allowed to correct it. If the error is of a nature that did not provide an unfair competitive advantage to the bidder and is not detrimental to the state, the contract may be awarded. Otherwise, the bid could be rejected. Specific comments included the following from Wisconsin:

> The foundation for decisions will continue to be based upon the State of Wisconsin Department of Transportation Standard Specification for Highway and Structure Construction Supplemental Specifications 2001 Edition sec. 102.6, 102.7, and 103.1 and the State of Wisconsin Statutes sec. 66.0901(5). Each situation is considered a unique situation and will be evaluated on a case-by-case basis as to how it relates to these specific specifications and statutes as well as any other applicable specification and statutes.
>
> As a supplement to the above evaluation, past decision on bidding issues will be revisited in order to ensure that a consistent and equitable decision will be made. Past situations where bidders have been allowed to withdraw their bids without loss of proposal guarantee include:
>
> a) A decimal point error made in a unit bid price where the bidder is able to provide clear and convincing evidence of both the mistake and the intended bid; and is able to demonstrate that the magnitude of the mistake would have significant impact on their ability to perform the work.
> b) Transposing or offsetting unit bid prices on multiple lines of the schedule of prices where the bidder is able to provide clear and convincing evidence of both the mistake and the intended bid; and is able to demonstrate that the magnitude of the mistake would have significant impact on their ability to perform the work.

Many other DOTs similarly referred to standard specifications that address the issue of errors in bids.

PRACTICES IN DATA SHARING OF LETTING PROGRAM INFORMATION

One premise of this synthesis is related to data sharing of letting program information among SHAs. This premise basically suggests that such data sharing would be beneficial for SHAs. The questionnaire attempted to assess the validity of this premise. As respondents indicated, data sharing of letting program information was not considered to be very important, with 50% viewing it as moderately important. Another 25% considered data sharing of letting

TABLE 11
DATA COLLECTED REGARDING LETTING PROGRAM MANAGEMENT

State	Types of Data Collected
Arkansas	The Department uses Trns•port software to collect and store data on a number of items, including the items requested in this question (bid unit prices, low bid unit prices, volume of work contracted by the contractor, or work remaining to be completed by project or contractor). In addition, our Construction Division uses Microsoft Access to link to data on the Trns•port file server to analyze work on the jobs by contractor, resident engineer, surety, and highway department district. This analysis includes time charges, percentage of time of current contract time used, contract amount, and percentage of payments (work complete) of current total contract amount.
Florida	Dollars let; number of projects let; type of funds used (both state and federal); type of construction (roadway, resurface, bridge, reconstruction, etc.); time and cost overruns for construction; number of projects planned compared to the actual number let for each year (Performance Report).
New York	Cost by county, fund type, work type, and letting agency; engineer's estimate versus low bid; contractor/vender owners and affiliates; each contractor/vender and their bid unit prices.
Washington	Unit bid prices of the three low bidders and the engineer's estimate for each contract is maintained on our unit bid analysis and standard item table; complete bid tabulations are maintained indefinitely; contractors' bid history; work in progress; engineer's estimate; all contract information related to cost is maintained in our Construction Contracts Information System—advertise date, bid opening date, award date, execution date, bid amount, final cost, etc.

TABLE 12
TYPICAL DATA ON PROGRAM MANAGEMENT THAT STATES WOULD LIKE TO COLLECT

State	Types of Data Desired from Other States
Iowa	Contractor performance.
Maine	We would like their (other states) web pages to include at least as much contractual, legal, specification, prequalification, and award data as ours, so we have access to an immediate and user-friendly source of information.
New York	Share responsibility decisions and pending issues with neighboring states; share letting and award data with neighboring states; specific issues and items are likely to vary over time. Current issues would include: contractor and subcontractor data. Data sharing between governmental units within the state is also desirable.
North Dakota	Procedures, methods, or programs they use to detect bid collusion, rigging, unbalancing, fraud, and so on.
Pennsylvania	Size of program—number of projects and dollars of program; and cycle times: bid opening to award, award to execution, and execution to notice to proceed.

program information as not important at all. The level of interest in this concept is not sufficiently high that SHAs would make a significant effort to exchange letting program management data, a finding confirmed as just over 50% of respondents reported that they do not share letting program data with other states. The remaining SHAs noted that they would share these data, but only if asked.

Most states collect some data in the letting program area, primarily related to bidding. Table 11 gives some examples of this type of data. In addition, most SHAs reported that they are collecting all the data they need for letting program management in the area of bidding, with one exception. Several states indicated that they would like more information on how to detect collusion in the bidding process. When asked what data SHAs would like from other states, agencies gave varied responses. Table 12 summarizes the responses of several states.

One area in which sharing letting schedule information might be useful is related to the bidding of large projects. For example, the Maryland DOT experienced problems with bids on the Woodrow Wilson Bridge superstructure contract. Only one bid was received, and that bid was 75% higher than the engineer's estimate. An independent review committee was organized to investigate this situation. The committee found that another major project was bid about the same time, making it difficult for the larger contractors to prepare simultaneous estimates for two mega projects ("Summary of Independent Review. . ." 2002). Thus, there was a lack of competition, resulting in higher bid prices ("Woodrow Wilson . . ." 2002). Sharing of letting information might have changed the timing of bids for the Maryland project, and better bids might have been obtained.

Most states have websites with some letting information available, such as bid tabulations of unit prices. These data are presented in a number of ways, such as by recent projects let (3-month period) or average bid prices on a statewide basis. Most bid prices correspond to detailed schedules of work items. Appendix B provides typical SHA website addresses where descriptions of this type of information can be located.

SUMMARY

The approach that SHAs follow to manage their letting programs does not appear to be documented by these agencies. However, of those SHAs that responded, most follow the generic process shown in Figure 6 in some form. Design completion is likely the most critical consideration when moving projects onto letting schedules. Matching funds to available projects is another key requirement to move projects onto letting schedules. Federal-aid projects must comply with FHWA requirements for project funding. States also closely monitor current cash flow and must ensure that sufficient funds are available before moving projects to letting schedules. Other constraints that can influence which projects are released to the letting schedule include SHA policy, geographical distribution of projects to districts or regions, and the priority set for the project based on the letting program and the STIP. The projected completion date of the PS&E final approval confirms the letting date on the schedule.

Letting schedules are constantly changing for many reasons. Managing the impact of change in letting schedules is critical to using available funds and to ensuring that projects are awarded for construction as planned. Regular meetings supported by project information systems that provide status on projects in the letting program pipeline are common techniques used to manage the letting program. Other factors can influence the final award of a project for construction, which when evaluated during bid analysis, may result in delaying the contract award. Finally, data sharing of letting information programs was determined to be a low priority item.

CHAPTER FIVE

CONCLUSIONS

This chapter provides key findings in the area of statewide highway letting program management. In general, statewide letting program management can be quite complex. Many agencies are involved, including both state highway agencies (SHAs) and external agencies. Within SHAs, many groups participate in various elements of the letting program process. It appears that few SHA personnel have a complete picture of how the total process is conducted within their agencies.

Funding plays a significant role in statewide letting program management. In particular, the Statewide Transportation Improvement Program (STIP) is fiscally constrained, which increases the burden of managing funds for projects within the STIP.

There appears to be no universal definition of a letting program, the time frame of a letting program (annual or other), nor when a letting program begins. The letting program is, however, closely linked with various stages within the SHA project development process.

A statewide highway letting program has two major components: (1) the development of the letting program and (2) the management of this program. There is currently no clear picture of how SHAs develop a letting program. As a result, documented processes were not uncovered that clearly describe this component. The preparation of an authorized program (6 to 12 years) is likely the first step in developing the letting program. A major component of the authorized program is the STIP. The project development process is closely tied to the STIP through project programming and the development of plan, specification, and estimate design documents for projects included in the STIP. Projects in the STIP are prioritized based on safety, existing facility condition, and other needs. This priority helps determine what projects move to the letting program or "pool of projects."

Processes exist for developing the SHA letting program; however, they are not usually well defined or formally documented. As with the development of the letting program, there appears to be no clear picture of how SHAs manage their letting programs. There are few documented processes that clearly and comprehensively describe this component. However, most SHAs indicated that the management of the letting program is an integral part of the project development process. Therefore, aspects of the management of the SHA letting program are included in various manuals on project development and other related procedures.

The manner in which projects are moved onto the letting schedule is closely tied to the availability of funds for projects in which plans, specifications, and estimates are complete or almost complete. This is the key requirement. Many other factors however can influence which projects are included on the letting schedule, such as environmental clearance, timely acquisition of right-of-way, geographical distribution of projects throughout the state, and policy and administrative issues.

Letting schedules are constantly changing for many reasons. Managing the impact of this change is critical. Several key sources of change are funding or cost, environmental clearance, right-of-way acquisition, and project scope. In these four areas, there are many causes of change in letting schedules. Many causes are related to the timeliness of completing design documents, obtaining approvals from various agencies, and changes in project scope. Actions taken by SHAs to mitigate the impact of change can be classified as reactive or preventive. Reactive actions may include rescheduling projects, shifting projects to other fiscal years, and moving other projects that are ready onto the letting schedule. Preventive actions may include improving the management of project schedules, monitoring cash flow, conducting periodic status meetings, and evaluating other important information needed to ensure that projects are let as scheduled.

There are other factors that influence the final award of construction contracts, such as nonresponsiveness of bidders, overruns and underruns of the engineer's estimate, bid errors, unbalanced bids, and evidence of collusion. SHAs have various policies and practices pertaining to how these factors are considered in awarding a bid. If a bid is rejected, then the project is most often rescheduled for a later letting. However, depending on the problem, such as significant overruns, the project might require some scope adjustment and redesign.

Finally, data sharing about letting programs is a low priority among SHAs. Very few SHAs believe this type of data sharing would be beneficial.

Overall, statewide highway letting program management practices are not well documented. The complexity and breadth of letting program management make it difficult for any one person or group within the SHA to understand the entire process. There appear to be a limited number of tools and techniques available to SHAs to aid them in statewide highway letting program management.

REFERENCES

Anderson, S.D. and D.J. Fisher, *NCHRP Report 390: Constructibility Review Process for Transportation Facilities*, Transportation Research Board, National Research Council, Washington, D.C., 1997, 96 pp.

"Financing Federal-Aid Highways," Publication FHWA-PL-99-015, Federal Highway Administration, U.S. Department of Transportation, Washington, D.C., 1999.

"Compass 2nd Quarter 2002," New Mexico Highway and Transportation Department, Santa Fe, 2002.

"Contract Administration Core Curriculum Participant's Manual and Reference Guide 2001," Federal Highway Administration, U.S. Department of Transportation, Washington, D.C. [Online]. Available: fhwa.dot.gov/programadmin/contracts/cor_I.htm [Feb. 2003].

Letting Capacity Enterprise-Wide Gap Closure, Draft 3—For Final Edit—LCS, Pennsylvania Department of Transportation, Harrisburg, July 2002.

Saag, J.B., *NCHRP Synthesis of Highway Practice 273: Project Development Methodologies for Reconstruction of Urban Freeways and Expressways*, Transportation Research Board, National Research Council, Washington, D.C., 1999, 45 pp.

Warne, T. and Maryland State Highway Administration, Summary of Independent Review Committee Findings Regarding the Woodrow Wilson Bridge Superstructure Contract, Mar. 1, 2002.

Woodrow Wilson Bridge Project, Bridge Superstructure Contract, BR-3: Review of the Engineer's Estimate versus the Single Bid, Feb. 28, 2002.

APPENDIX A

Survey Questionnaire

NCHRP PROJECT 20-5 SYNTHESIS

Topic 33-09

STATEWIDE HIGHWAY LETTING PROGRAM MANAGEMENT

QUESTIONNAIRE

PURPOSE OF THE SYNTHESIS

State highway agencies (SHAs) have established processes, with varying degrees of complexity, to manage annual highway letting programs. These processes include tools and techniques used to schedule lettings and award transportation projects in compliance with certain established criteria. However, the letting process can be volatile, which can cause changes in planned letting schedules. Understanding the causes behind change and how change is managed with respect to the letting process is important.

The management of the annual highway letting programs is a critical component of successful project delivery and the achievement of metropolitan planning organization, state, and national goals. The purpose of this synthesis is to summarize available information and document how SHAs manage their annual letting programs.

RESPONDING AGENCY INFORMATION

Please complete the following request for information to aid in processing this questionnaire:

Agency: ____

Address: ____

____

City: ______________________ State: ______________ Zip: __________

Questionnaire completed by: ______________________________

Current position/title: ______________________________

Date: ______________________ E-mail: ______________________

Telephone: ______________________ Fax: ______________________

Agency contact (if different from above): ______________________

Telephone: ______________________ E-mail: ______________________

PLEASE RETURN THE COMPLETED QUESTIONNAIRE BY AUGUST 2, 2002.

To: Stuart D. Anderson
Department of Civil Engineering
3136 TAMU
Texas A&M University
College Station, Texas 77843-3136

Telephone: 979-845-2407
Fax: 979-845-6554
E-mail: s-anderson5@tamu.edu

Please contact Stuart Anderson if you have questions.

OBJECTIVES

This synthesis will identify the processes used by SHAs for highway letting program management. Specific objectives of the synthesis are:

- Describe letting program management structures,
- Evaluate how change influences the annual letting program process,
- Identify issues with contract award, and
- Assess data sharing initiatives.

SYNTHESIS LIMITS

The scope of this synthesis covers **the preparation and administration of the annual letting program through construction contract award**.

INSTRUCTIONS

Please be concise with your answers. Because many questions are open-ended, follow-up telephone interviews will likely be necessary to confirm or enhance the understanding of the responses. Please be sure you provide us with a contact person for this purpose.

Please enclose any information you believe is relevant to the answers provided in the questionnaire, including applicable procedures, policies, or other information that might be of interest to other state highway agencies.

THANK YOU IN ADVANCE FOR YOUR HELP AND COOPERATION WITH THIS PROJECT!

I. PROJECT DEVELOPMENT AND LETTING PROGRAM

1. List the **main phases** (or steps) in your agency's project development process (from project need identification to construction contract award).

2. At what **phase** (or step) in the project development process (question 1) is a project included in the annual letting program?

3. Are state/local-funded projects administered differently from federal-aid projects?

   Yes _____ No_____

   If yes, briefly describe the differences.

4. What process is used to ensure that there are sufficient projects to fully utilize available funds?

II. LETTING PROGRAM

5. Please describe the process your agency follows for developing your annual letting program.

   Is this process documented? Yes ____ No____

   If yes, is this document located on your website? Yes ____ No____
   If yes, what is the website address and how can the document be located (i.e., under what categories, sections, etc.)?

   If no, can you send a hard copy or advise how to obtain a hard copy?

6. Who in your agency is responsible for managing the annual letting program? If different then the name(s) on page 1, please provide this person's name, address, telephone number, and e-mail address below.

7. How does your agency select the projects to be included in the annual letting program?

8. How frequently are projects let for construction bidding? (Check dominant one.)

Monthly ______
Bi-monthly ______
Quarterly ______
Semi-annually ______
Other (please specify) ________________ ______

9. How does your agency select projects for **specific** lettings?

10. What letting program steps (or activities) are mandated by state law? (Check those that apply.)

a. Bid advertising ______
b. Bid requirements (length of time, location, etc.) ______
c. Treatment of out-of-state contractors ______
d. Prequalification of contractors ______
e. Electronic submission (e-mail) ______
f. Internet submission through website ______
g. Other—Please list ______________________________
____
____

What are requirements for those steps (or activities) that are affected by state law?

a.

b.

c.

d.

e.

f.

g.

11. What letting program steps (or activities) are mandated by federal law? (Check those that apply.)

a. Bid advertising ______
b. Bid requirements (length of time, location, etc.) ______
c. Treatment of out-of-state contractors ______
d. Prequalification of contractors ______
e. Electronic submission (e-mail) ______
f. Internet submission through website ______
g. Other—Please List ______________________________
____
____

What are requirements for those steps (or activities) that are affected by federal law?

a.

b.

c.

d.

e.

f.

g.

12. What other agencies are involved in the annual letting program process? What is their role?

III. VOLATILITY IN LETTING PROGRAM

13a. Identify the top five major factors that have resulted in changes to the projects included in your annual letting program.

1.

2.

3.

4.

5.

13b. What were the main causes of these changes?

1.

2.

3.

4.

5.

13c. What actions did you take when trying to manage the impacts that resulted from these changes?

1.

2.

3.

4.

5.

13d. How did these changes impact the annual letting program?

1.

2.

3.

4.

5.

IV. CONTRACT AWARD CONSIDERATIONS

14a. Please complete the following Table

Construction Procurement	Used by Agency (Check)	Impact on Letting Process
Traditional		
Design–build		
Prequalify		
Stand-by		
Emergency		
Advanced procurement		
Work order		
Job order contracting		

15. Does your agency always award the contract to the low bidder if the contract is to be awarded?

Yes _____ No ______ If no, what are reasons for not doing so?

16. Describe your agency's bidding process.

17. Does your agency have criteria for rejecting bids that overrun the engineer's estimate?

    Yes ______ No ______ If yes, please list them.

18. Does your agency analyze bids that appear to be unbalanced?

    Yes ______ No ______ If yes, briefly describe how this analysis is performed (especially note if a computer program is used).

19. Does your agency have a procedure to identify whether or not collusion is possibly present in the bidding process?

    Yes ______ No ______ If yes, briefly describe how this procedure is performed.

20. How does your agency determine if a bid is not complete and/or non-responsive?

21. What policies do you follow if there are errors in bids?

22. Does your agency permit submission of electronic bids? Yes ______ No ______

    If yes, how do you maintain security?

23. Does your agency conduct pre-bid conferences? Yes ______ No ______

    If yes, what are the conditions/criteria for conducting this conference?

24. Is 100 percent clearance of the following areas required before letting a project?

Utilities Yes ___ No ____ If no, what is typical percent of area cleared _____%
Right-of-way Yes ___ No ____ If no, what is typical percent of area cleared _____%
Permits Yes ___ No ____ If no, what is typical percent of area cleared _____%
Other please specify:

____
____
____
____
____

V. DATA SHARING

25. What types of data does your agency collect with respect to your annual letting program (for example, bid unit prices, low bid unit prices, volume of work contracted by the contractor, or work remaining to be completed by project or contractor)?

26. Are there data you would like to collect but are not currently collecting?

Yes _____ No _____If yes, what data would you like to collect?

27. Does your agency share data with other state highway agencies with respect to letting program information?

Yes _____ No _____ If yes, what types of data do you share? With which states do you share these data?

28. What data would you want other SHAs to share with you?

29. How important would data sharing of letting program information be to your agency?

Very important _____
Important _____
Moderately important _____
Not Important _____

30. Does your agency use any of the following AASHTO Trns•port System Programs?

Cost Estimating System (CES)	Yes ____	No ____
Proposal and Estimating System (PES)	Yes ____	No ____
Letting and Awards System (LAS)	Yes ____	No ____
Expedite (Electronic Bid Letting)	Yes ____	No ____

Please provide comments on the use of any of these systems.

THANK YOU AGAIN FOR PARTICIPATING IN THIS SYNTHESIS STUDY!

APPENDIX B

State Highway Agency Websites

This appendix presents illustrations from selected SHA websites that provide information relevant to the subject matter discussed and referenced in the body of the synthesis. In general, most state highway agencies have similar information on their websites. Because websites are continually updated and modified, the specific information cited in this appendix may change over time.

Arkansas State Highway and Transportation Department

The general website address for the Arkansas State Highway and Transportation Department is ArkansasHighways.com. The user can click on "Contracts" and then "Construction Contract Information" to find the latest STIP for a 3-year period. By clicking on the STIP icon, the user can access a 1-page summary of the current STIP. Other information can be reviewed in this same area, such as a letting schedule, which in this case gives the dates (month/day). Bid data can also be accessed in this same area.

Georgia Department of Transportation

The Georgia DOT provided the following website address for information: www.dot.state.ga.us/dot/administration/financialmgmt/index.shtml. This specific address, under the Office of Financial Management, offers information on the Project Selection Process and the Federal Funding Process. This website also contains information about the STIP development process and the Plan Development Process. Sort descriptions of these areas are also provided. Additional information can be retrieved by clicking on "Plans and Programs." Finally, by clicking on "Business Opportunities," the user can obtain letting schedule information.

Pennsylvania Department of Transportation

The general website address for the Pennsylvania DOT is www.dot.state.pa.us. The user can click on "Doing Business with PENNDOT," then "Select EBS/ECMS," and then click on "ECMS." In the ECMS section, the user can find letting information by clicking on "Electronic Bidding" and then "Letting Schedule." Successfully click on "References," "Highway Related Pubs," and "Publication 10—Transportation Project Development Process" to find out about program management. If the user clicks on "PennDOT Systems" and "MPMS," information can be found about the STIP.

Washington State Department of Transportation

The Washington State DOT provided the following website address for information on its Capital Improvement Program Process (CIPP): www.wsdot.wa.gov/projects/cipp/. This website contains a description and discussion of the Capital Improvement and Preservation Program. Information relevant to the synthesis can be found by clicking on different icons such as "Program Summary" for highways.

APPENDIX C

Telephone Interview Protocol

NCHRP PROJECT 20-5 TOPIC 33-09

STATEWIDE HIGHWAY LETTING PROGRAM MANAGEMENT

Telephone Interview Protocol

Purpose

To gain a better understanding of how state highway agencies develop the letting program and then manage this program.

Target States

Small—Connecticut, Delaware, and Maryland
Medium—Arkansas, Kansas, and Washington
Large—California, New York, Pennsylvania, and Texas

Interview Framework

Generic process flowchart with questions based on flowchart and state agency responses from written questionnaire [see overall flowchart (Figure 3) in the report and more detail on portion of overall flow chart (Figure 6) in the report].

Interview Setup

Make contact and obtain agreement for interview,
Send framework graphic and potential question areas, and
Set interview time.

Interviews

Provide brief overview of generic process;
Clarify key definitions—annual letting program, etc.; and
Ask questions.

Questions

General

1. Does the generic process flowchart in Figure 3 *generally* fit the approach your state follows to develop and manage your letting program? If not, what part of the process would be different or what part would you modify to fit your state's approach? Explain.

Process for Developing Letting Program

1. Discuss, in general, how you develop your State Transportation Improvement Program (STIP) (see Figure 3).
2. How do you prioritize projects included in your STIP? What factors do you consider in this prioritization process?
3. Approximately how many projects would be included in your STIP? Approximately how many total projects are under development at any one time (including those in the STIP)?
4. What term do you use to describe the "pool of projects" shown on Figure 3?
5. What key factors would determine when a project enters the pool of projects that collectively comprise a letting program; for example, status of plans, specifications, and estimates percent complete and so on (see Figure 3)?

Management of Letting Program

1. How do you track which projects are ready to go into the pool of projects? What tools do you use for this tracking?
2. What factors are key to moving a project from the "pool of projects" to the letting schedule (see Figure 3)?
3. What is the typical time period that is covered on your letting schedule (e.g., projects let within 6 months, one year)?
4. How do you prioritize projects for your letting schedule?
5. How do available funds influence when projects are included on the letting schedule? What tools do you use to

determine the source and amount of funds available for these projects?

6. What other factors do you consider when moving projects onto your letting schedule, such as bidding volume in a given geographical region; availability of contractors; type of projects; or district, division, region requests?

7. What techniques and tools do you use to monitor the letting schedule to ensure that a specific project is let according to this schedule?

8. What will change your letting schedule (e.g., environmental approval, funding modifications, cash flow restrictions, political directives, etc.)? How do you manage this change? What tools do you use?

APPENDIX D

NMDOT Letting Schedule Tracking Illustrations

Result 15: Stable letting schedule to better manage limited resources in order to meet more of our customer's needs.

Result Driver: Charlie Trujillo

The department's Letting Schedule is an important tool to ensure full utilization of both Federal and State funding; it reflects the stability of the department's Statewide Transportation Improvement Program and its long-term plans. The Letting Schedule serves as a managing/planning guide for Engineering/Design resource needs; and it also provides mid-range project letting information to the general public, as well as highway construction contractors. A stable Letting Schedule is an indicator of the department's ability and commitment to manage the funding and human resources necessary to accomplish the department's Vision.

15a. Percentage of projects let as scheduled three months previously.

Measurement Driver: Chris Ortega
Measurement Team: Kenny Luján Sr.

Revised: Friday, July 12, 2002

Purpose
This measurement provides one indication of how well the Department is doing in planning and managing the project development process and projecting anticipated project letting dates. An accurate letting schedule allows the Department to efficiently utilize design resources and contractors to make financial management decisions to plan their resources.

Data Collection
The Letting Schedule lists the highway construction projects to be let each month by the department for the next 12 months. The total number of projects let on schedule, (as projected) for a quarter is divided by the number of projects listed in the schedule for that quarter.

Measurement
A higher percent indicates increased accuracy of projected letting dates in the Letting Schedule.

Improvement
By increasing the accuracy of the Letting Schedule, the department is improving its planning and performance to better apply our resources and to better serve our customers.

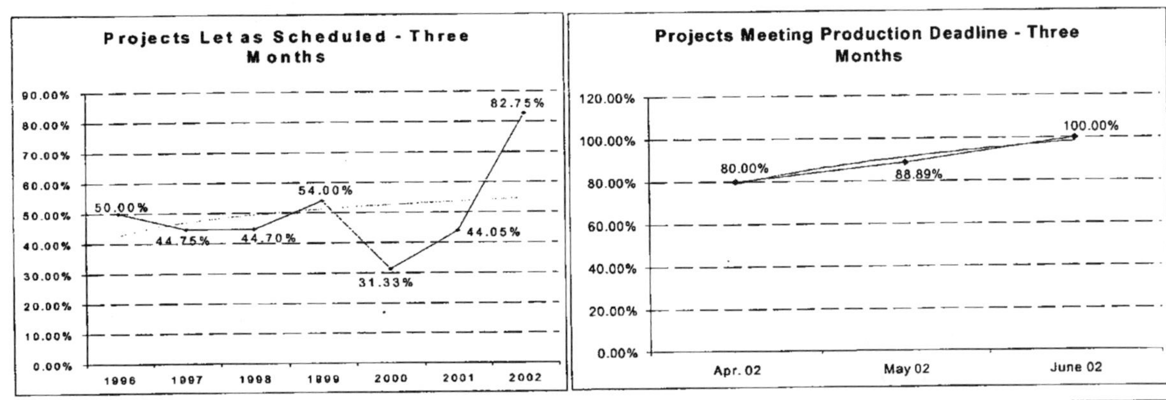

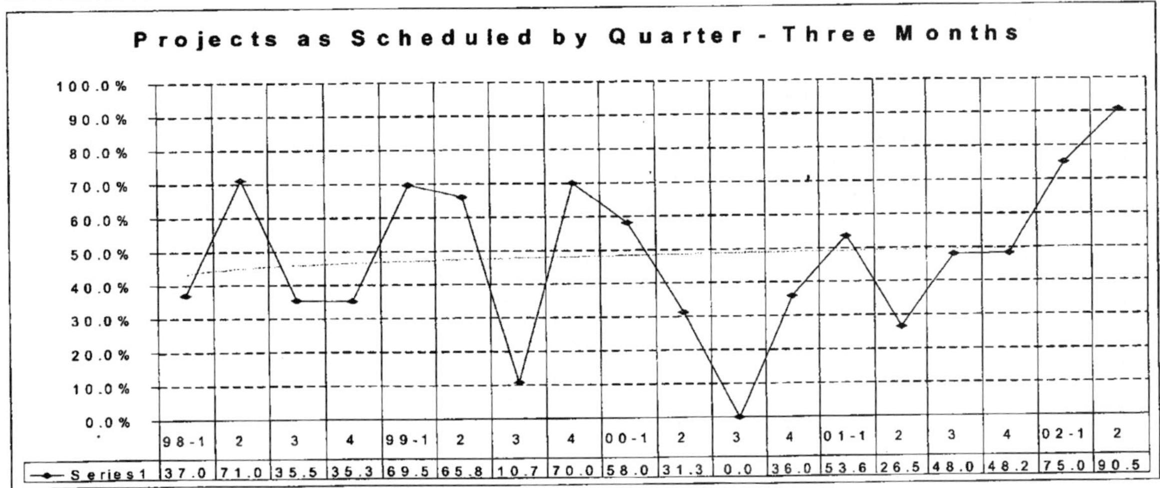

15b. Percentage of projects let as scheduled six months previously.

Measurement Driver: Chris Ortega
Measurement Team: Kenny Luján Sr.

Revised: Friday, July 12, 2002

Purpose
This measurement provides one indication of how well the Department is doing in planning and managing the project development process and projecting anticipated project letting dates. An accurate letting schedule allows the Department to efficiently utilize design resources and contractors to make financial management decisions to plan their resources.

Data Collection
The Letting Schedule lists the highway construction projects to be let each month by the department for the next 12 months. The total number of projects let on schedule, (as projected) for six months is divided by the number of projects listed in the schedule for that six month period.

Measurement
A higher percent indicates increased accuracy of projected letting dates in the Letting Schedule.

Improvement
By increasing the accuracy of the Letting Schedule, the department is improving its planning and performance to better apply our resources and to better serve our customers.

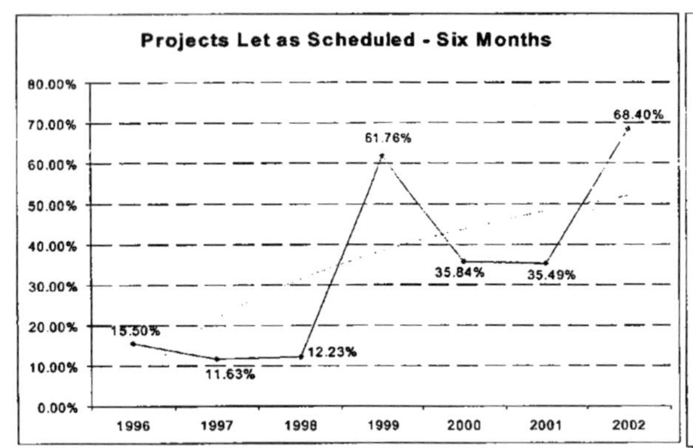

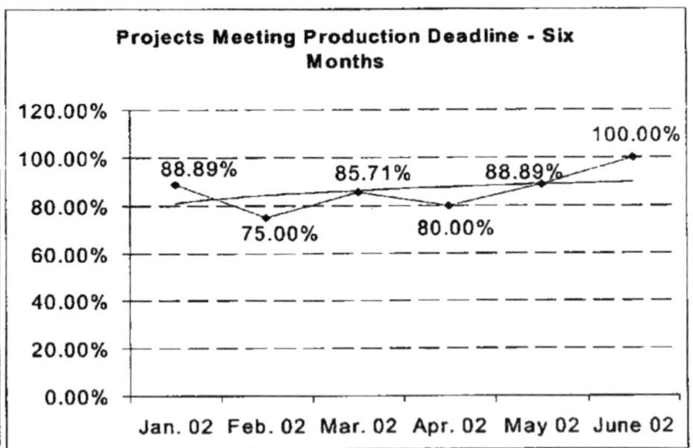

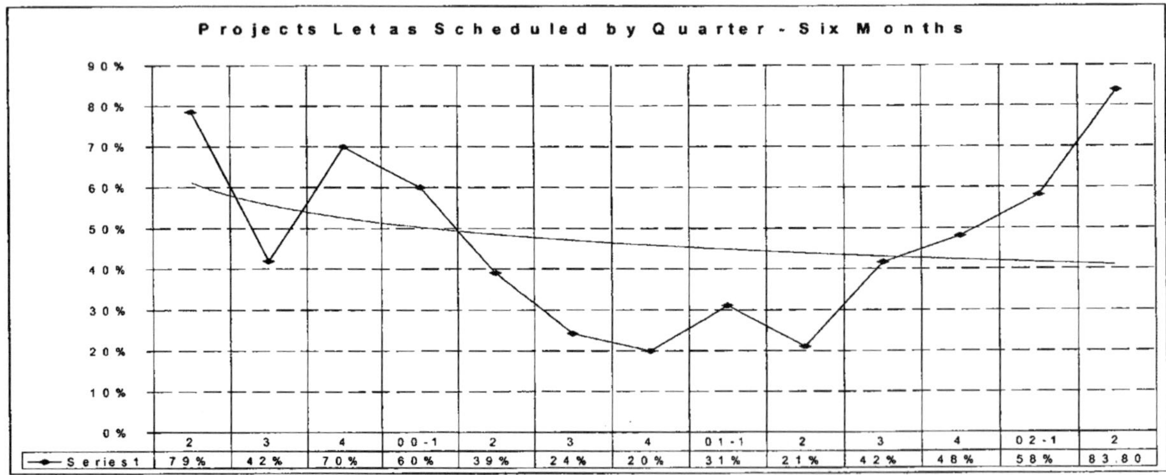

15c. Percentage of projects let as scheduled one year previously.

Measurement Driver: Chris Ortega
Measurement Team: Kenny Luján Sr.

Revised: Friday, July 12, 2002

Purpose
This quantitative measurement provides a clear indication of how well the department is planning and managing its limited resources. A stable letting schedule allows contractors to make financial management decisions and plan their resources. Contractors depend on a monthly letting schedule to continue their working operations.

Data Collection
The Letting Schedule lists the highway construction projects to be let each month by the department for the next 12 months. The total number of projects let on schedule for a year is divided by the number of projects listed in the Letting Schedule for that year.

Measurement
A higher measurement indicates increased accuracy of the Letting Schedule.
Improvement
By increasing the accuracy of the Letting Schedule, the department is improving its planning and management performance to better serve its customers.

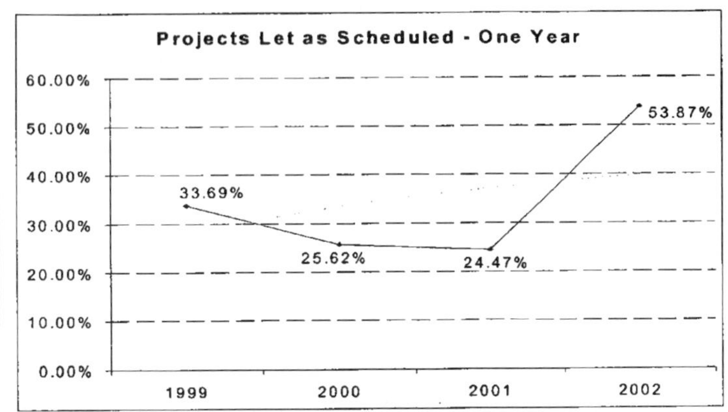

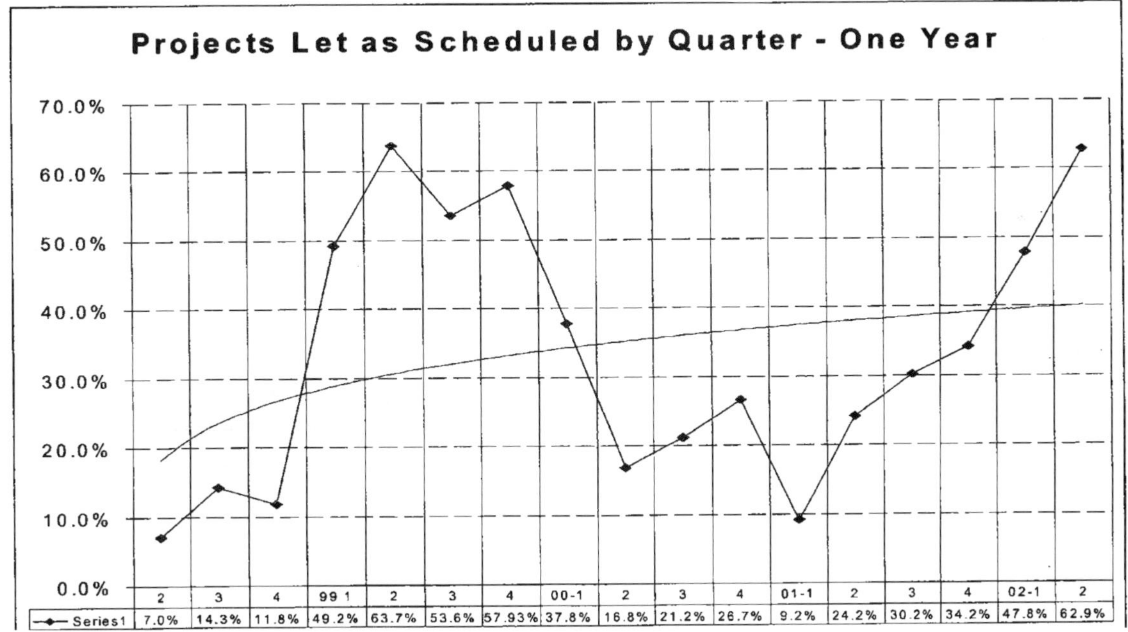

	2	3	4	99 1	2	3	4	00-1	2	3	4	01-1	2	3	4	02-1	2
Series1	7.0%	14.3%	11.8%	49.2%	63.7%	53.6%	57.93%	37.8%	16.8%	21.2%	26.7%	9.2%	24.2%	30.2%	34.2%	47.8%	62.9%

15d. Federal-Aid Program Federal Limitation, Cumulative Average Limitation, and Cumulative Obligation.

Measurement Driver: Chris Ortega
Measurement Team: Arthur R. Gurulé

Revised: 7/12/02

Purpose
Federal funds are a major component of the Department's construction program. The Department authorizes Federal funds for project lettings and construction related activities. This measurement reflects the ability of the Department to commit and meet the Federal-Aid Program. Federal funds are authorized at least a month before a project is let.

Data Collection
The letting schedule is a tool used to time the letting of construction projects and utilize the Department's Federal-Aid Obligation Authority. The Department's goal is to meet each year's Federal-Aid Program.

To measure performance, data is collected from the Federal Obligation Report. The value reported is the monthly Cumulative Obligation (CO), which is the value of projects that have been authorized by FHWA for Federal reimbursement.

Once a Transportation Bill becomes law, FHWA sets Fiscal Year Federal Limitations, which are the maximum Federal funds available for reim-bursement to the Department. The Cumulative Average Limitation (CAL) is a straight-line spread of the FY 02 Federal Limitation.

Measurement
The CO measurement provides a clear indication of how the Department is obligating Federal funds. Also, the department is able to adjust the monthly letting schedule to meet the CAL.

The CO line is flatter at the beginning of the Fiscal Year due to the Department only having Continuing Resolutions to obligate until late December 2001. Although the CO line has steepened, it is still below the CAL partially due to the Debt Service not being fully obligated at this point. It is expected that by the end of the Fiscal Year all the Federal-Aid Limitation will be obligated. The Department met the goal using all of the Federal-Aid Limitation in FY 2001.

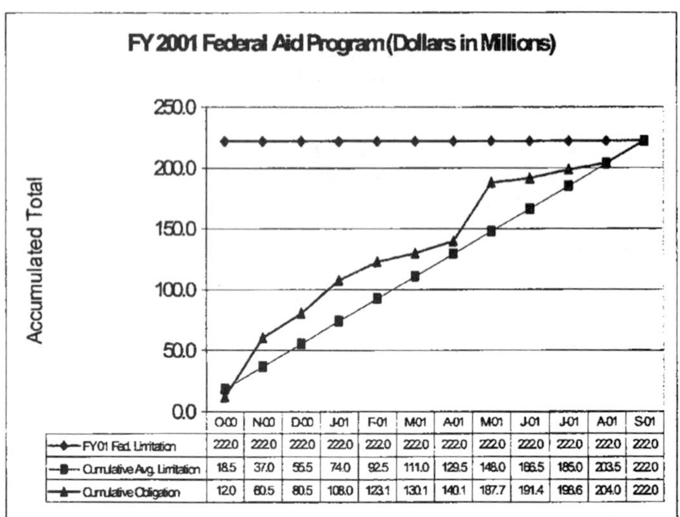

	O-00	N-00	D-00	J-01	F-01	M-01	A-01	M-01	J-01	J-01	A-01	S-01
FY01 Fed. Limitation	222.0	222.0	222.0	222.0	222.0	222.0	222.0	222.0	222.0	222.0	222.0	222.0
Cumulative Avg. Limitation	18.5	37.0	55.5	74.0	92.5	111.0	129.5	148.0	166.5	185.0	203.5	222.0
Cumulative Obligation	12.0	60.5	80.5	108.0	123.1	130.1	140.1	187.7	191.4	198.6	204.0	222.0

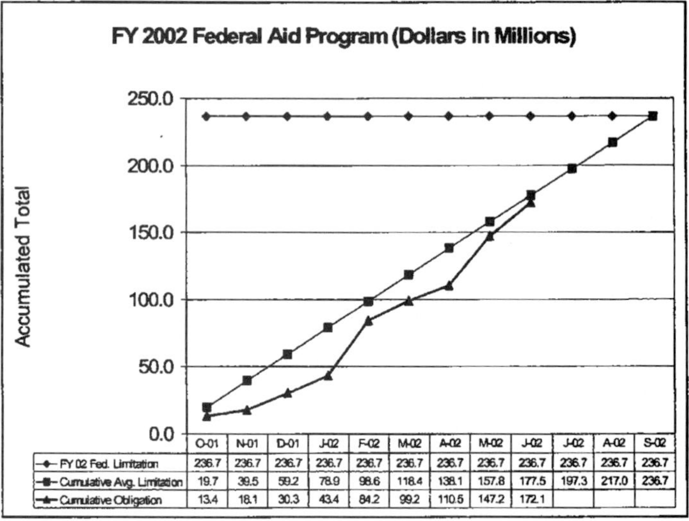

	O-01	N-01	D-01	J-02	F-02	M-02	A-02	M-02	J-02	J-02	A-02	S-02
FY 02 Fed. Limitation	236.7	236.7	236.7	236.7	236.7	236.7	236.7	236.7	236.7	236.7	236.7	236.7
Cumulative Avg. Limitation	19.7	39.5	59.2	78.9	98.6	118.4	138.1	157.8	177.5	197.3	217.0	236.7
Cumulative Obligation	13.4	18.1	30.3	43.4	84.2	99.2	110.5	147.2	172.1			

15e. State Program: Cumulative Average Budget and Cumulative Obligation.

Measurement Driver: Chris Ortega
Measurement Team: Arthur R. Gurulé

Revised 7/12/02

Purpose

The Department's letting schedule is an important tool to ensure full utilization of the State Highway Program. The letting schedule is constrained by the available budget and can be adjusted to match cash flow needs. It reflects the ability of the department to commit and meet the state program within available financial resources.

Data Collection

The letting schedule is a tool used to time the letting of construction projects in order to maximize the Department's State Highway Program. The Department's goal is to meet each year's State Program. To measure performance, data is collected from the projects that have been let and encumbered, plus activity from prior year projects that affects the current budget.

Measurement

The Cumulative Obligation measurement provides a clear indication of how the Department is doing managing the State Program. The Department is able to adjust the monthly letting schedule to match available budget and cash resources.

Throughout the fiscal year the Department generated budget / cash savings and applied the savings to the 100% State program in order to contract pavement preservation projects. The letting schedule was modified to utilize the budget / cash savings and the projects were let in May and encumbered in June. Several projects were also awarded through existing State Price Agreements.

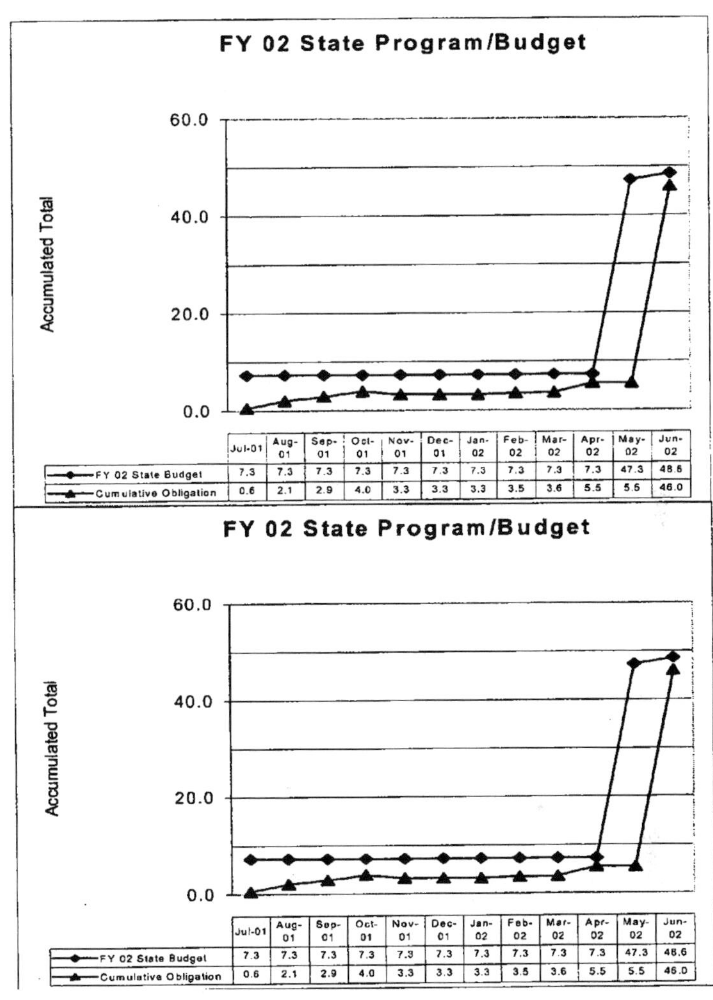

	Jul-01	Aug-01	Sep-01	Oct-01	Nov-01	Dec-01	Jan-02	Feb-02	Mar-02	Apr-02	May-02	Jun-02
FY 02 State Budget	7.3	7.3	7.3	7.3	7.3	7.3	7.3	7.3	7.3	7.3	47.3	48.6
Cumulative Obligation	0.6	2.1	2.9	4.0	3.3	3.3	3.3	3.5	3.6	5.5	5.5	46.0

	Jul-01	Aug-01	Sep-01	Oct-01	Nov-01	Dec-01	Jan-02	Feb-02	Mar-02	Apr-02	May-02	Jun-02
FY 02 State Budget	7.3	7.3	7.3	7.3	7.3	7.3	7.3	7.3	7.3	7.3	47.3	48.6
Cumulative Obligation	0.6	2.1	2.9	4.0	3.3	3.3	3.3	3.5	3.6	5.5	5.5	46.0